NATIONAL GEOGRAPHIC

World Cultures and Geography

EUROPE

GEO

Go interactive with **myNGconnect.com**

Acknowledgments

Grateful acknowledgment is given to the authors, artists, photographers, museums, publishers, and agents for permission to reprint copyrighted material. Every effort has been made to secure the appropriate permission. If any omissions have been made or if corrections are required, please contact the Publisher.

Photographic Credits

Front Cover: © Roger Ressmeyer/Corbis
Back Cover: © Neale Clark/Robert Harding World Imagery/Getty Images

Acknowledgments and credits continued on page RB144.

Printed in the USA
RR Donnelley
Menasha, WI

ISBN: 978-07362-9027-2

13 14 15 16 17 18 19 20

10 9 8 7 6 5 4 3 2

World Cultures and Geography OVERVIEW

STUDENT EDITION UNITS	COMPREHENSIVE (SURVEY)	EASTERN HEMISPHERE	WESTERN HEMISPHERE
THE ESSENTIALS OF GEOGRAPHY	•	•	•
NORTH AMERICA	•		•
Central America & the Caribbean	•		•
SOUTH AMERICA	•		•
EUROPE	•	•	•
RUSSIA & THE EURASIAN REPUBLICS	•	•	•
Sub-Saharan Africa	•	•	
SOUTHWEST ASIA & NORTH AFRICA	•	•	
South Asia	•	•	
East Asia	•	•	
Southeast Asia	•	•	
AUSTRALIA, THE PACIFIC REALM & ANTARCTICA	•	•	

EUROPE

TEACHER'S EDITION

CONSULTANTS AND REVIEWERS

Program Consultants

Peggy Altoff
District Coordinator
Past President, NCSS

Mark H. Bockenhauer
Professor of Geography,
St. Norbert College

Andrew J. Milson
Professor of Social Science Education
and Geography, University of Texas (Arlington)

David W. Moore
Professor of Education,
Arizona State University (Phoenix)

Janet Smith
Associate Professor of Geography,
Shippensburg University

Michael W. Smith
Professor, Department of Curriculum,
Instruction, and Technology in Education,
Temple University

Teacher Reviewers

Kayce Forbes
Deerpark Middle School
Austin, Texas

Michael Koren
Maple Dale School
Fox Point, Wisconsin

Patricia Lewis
Humble Middle School
Humble, Texas

Julie Mitchell
Lake Forest Middle School
Cleveland, Tennessee

Linda O'Connor
Northeast Independent School District
San Antonio, Texas

Leah Perry
Exploris Middle School
Raleigh, North Carolina

Robert Poirier
North Andover Middle School
North Andover, Massachusetts

Heather Rountree
Bedford Heights Elementary
Bedford, Texas

Erin Stevens
Quabbin Regional Middle/High School
Barre, Massachusetts

Beth Tipper
Crofton Middle School
Crofton, Maryland

Mary Trichel
Atascocita Middle School
Humble, Texas

Andrea Wallenbeck
Exploris Middle School
Raleigh, North Carolina

Reviewers of Religious Content

Charles Haynes
First Amendment Center
Washington, D.C.

Shabbir Mansuri
Institute on Religion and
Civic Values
Fountain Valley, California

Susan Mogull
Institute for Curriculum Reform
San Francisco, California

Raka Ray
Chair, Center for South Asia Studies
University of California
(Berkeley)

NATIONAL GEOGRAPHIC EXPLORERS, FELLOWS, AND GRANTEES

Greg Anderson
National Geographic Fellow

Katey Walter Anthony
National Geographic Emerging Explorer

Katy Croff Bell
National Geographic Emerging Explorer

Alexandra Cousteau
National Geographic Emerging Explorer

Thomas Taha Rassam (TH) Culhane
National Geographic Emerging Explorer

Jenny Daltry
National Geographic Emerging Explorer

Wade Davis
National Geographic Explorer-in-Residence

Sylvia Earle
National Geographic Explorer-in-Residence

Grace Gobbo
National Geographic Emerging Explorer

Beverly Goodman
National Geographic Emerging Explorer

David Harrison
National Geographic Fellow

Kristofer Helgen
National Geographic Emerging Explorer

Fredrik Hiebert
National Geographic Fellow

Zeb Hogan
National Geographic Fellow

Shafqat Hussain
National Geographic Emerging Explorer

Beverly Joubert
National Geographic Explorer-in-Residence

Dereck Joubert
National Geographic Explorer-in-Residence

Albert Lin
National Geographic Emerging Explorer

Elizabeth Kapu'uwailani Lindsey
National Geographic Fellow

Kakenya Ntaiya
National Geographic Emerging Explorer

Enric Sala
National Geographic Explorer-in-Residence

Kira Salak
National Geographic Emerging Explorer

Katsufumi Sato
National Geographic Emerging Explorer

Spencer Wells
National Geographic Explorer-in-Residence

Best Practices
For ACTIVE Teaching

To bring best practices into your classroom, choose from the following technology components and instructional routines and make them part of your daily instruction. Many of these practices are also built directly into the instruction in your Teacher's Edition.

PROGRAM TECHNOLOGY

World Cultures and Geography provides a variety of technology to make your job easier and to help students become motivated, independent learners. Use the online components listed below to supplement print resources or to create an entirely digital learning environment.

STUDENT COMPONENTS	TEACHER COMPONENTS
Student eEdition	Teacher's eEdition
Interactive Map Tool	Core Content Presentations
Digital Library	Online Lesson Planner
Magazine Maker	Assessment
Connect to NG	Interactive Whiteboard GeoActivities
Maps and Graphs	Guided Writing
GeoJournal	Teacher Resources
Student Resources	

All digital resources and more information about them are available at **myNGconnect.com**. In addition, see the next page for specific instructions on how to use the **Interactive Map Tool**—a technology component that plays an integral part in the Teacher's Edition instruction.

Interactive Map Tool

The **Interactive Map Tool** is an online mapmaker that allows students to draw and add labels and data layers to a map. For general use, follow the instructions below to navigate the map tool.

1 Access the tool at **myNGconnect.com**.

2 Select a region to explore. Click on the "Region" menu to select a continent and the "Country" menu to select a specific country, or simply click and drag the map to maneuver it to a specific location. You can also use the slider bar to zoom in and out.

3 Choose the type of map you wish to view by clicking on the "Map Mode" menu. Options include terrain, topographic, satellite, street, National Geographic, and outline. The Outline mode allows you to click on a specific country and access a separate outline map of that country with features that you can turn on or off.

4 Click on the THEMES tab to display categories of data layers that can be overlaid on the map. Availability of these data layers varies depending on the zoom level of the map. Each data layer also has a legend and a transparency control.

5 Click on the DRAWING TOOLS tab to display a variety of tools that you can use to draw on the map. Most tools allow you to adjust the outline and fill colors, line width, and transparency. The tab can also be selected by clicking on the "Draw" icon.

6 Click on the MARKERS tab to display a variety of markers that you can drag and drop on the map. You can set the markers at three different sizes. The MARKERS tab can also be selected by clicking on the "Markers" icon.

7 To easily measure distances on the map, click on the "Measure" (ruler) icon. Click once on the map to start measuring, move the pointer to another spot, and click again to stop. The resulting line will display the distance in either kilometers or miles based on your selection from the dropdown menu.

Interactive Map Tool

COOPERATIVE LEARNING

Cooperative learning strategies transform today's classroom diversity into a vital resource for promoting students' acquisition of both challenging academic content and language. These strategies promote active engagement and social motivation for all students.

STRUCTURE & GRAPHIC	DESCRIPTION	BENEFITS & PURPOSES
CORNERS A strongly agree B disagree C agree D strongly disagree	• Corners of the classroom are designated for focused discussion of four aspects of a topic. • Students individually think and write about the topic for a short time. • Students group into the corner of their choice and discuss the topic. • At least one student from each corner shares about the corner discussion.	• By "voting" with their feet, students literally take a position about a topic. • Focused discussion develops deeper thought about a topic. • Students experience many valid points of view about a topic.
FISHBOWL	• Part of the class sits in a close circle facing inward; the other part of the class sits in a larger circle around them. • Students on the inside discuss a topic while those outside listen for new information and/or evaluate the discussion according to pre-established criteria. • Groups reverse positions.	• Focused listening enhances knowledge acquisition and listening skills. • Peer evaluation supports development of specific discussion skills. • Identification of criteria for evaluation promotes self-monitoring.
INSIDE-OUTSIDE CIRCLE	• Students stand in concentric circles facing each other. • Students in the outside circle ask questions; those inside answer. • On a signal, students rotate to create new partnerships. • On another signal, students trade inside/outside roles.	• Talking one-on-one with a variety of partners gives risk-free practice in speaking skills. • Interactions can be structured to focus on specific speaking skills. • Students practice both speaking and active listening.
JIGSAW Expert Group 1 — A's Expert Group 2 — B's Expert Group 3 — C's Expert Group 4 — D's	• Students are grouped evenly into "expert" groups. • Expert groups study one topic or aspect of a topic in depth. • Students regroup so that each new group has at least one member from each expert group. • Experts report on their study. Other students learn from the experts.	• Becoming an expert provides in-depth understanding in one aspect of study. • Learning from peers provides breadth of understanding of over-arching concepts.

STRUCTURE & GRAPHIC	DESCRIPTION	BENEFITS & PURPOSES
NUMBERED HEADS 1 — Think Time — 2 4 — 3 1 — Talk Time — 2 4 — 3 Share 2's Time	• Students number off within each group. • Teacher prompts or gives a directive. • Students think individually about the topic. • Groups discuss the topic so that any member of the group can report for the group. • Teacher calls a number and the student with that number reports for the group.	• Group discussion of topics provides each student with language and concept understanding. • Random recitation provides an opportunity for evaluation of both individual and group progress.
ROUNDTABLE 4 — 1 3 — 2	• Students sit around tables in groups of four. • Teacher asks a question with many possible answers. • Each student around the table answers the question a different way.	• Encouraging elaboration creates appreciation for diversity of opinion and thought. • Eliciting multiple answers enhances language fluency.
TEAM WORD WEBBING A D — B C	• Teams of students sit around a large piece of paper. Each team member has a different colored marker. • Teacher assigns a topic for a Word Web. • Each student adds to the part of the web nearest to him/her. • On a signal, students rotate the paper and each student adds to the nearest part again.	• Individual input to a group product ensures participation by all students. • Shifting point of view supports both broad and in-depth understanding of concepts.
THINK, PAIR, SHARE Think A B Pair A B Share A B	• Students think about a topic suggested by the teacher. • Pairs discuss the topic. • Students individually share information with the class.	• The opportunity for self-talk during the individual think time allows the student to formulate thoughts before speaking. • Discussion with a partner reduces performance anxiety and enhances understanding.
THREE-STEP INTERVIEW A — 1 / 2 — B 3 GROUP	• Students form pairs. • Student A interviews Student B about a topic. • Partners reverse roles. • Student A shares with the class information from Student B; then B shares information from Student A.	• Interviewing supports language acquisition by providing scripts for expression. • Responding provides opportunities for structured self-expression.

UNIT 2 EUROPE

TECHTREK

myNGconnect.com

Digital Library
Unit 2 GeoVideo
Introduce Europe

Explorer Video Clip
Enric Sala, Marine Ecologist
National Geographic Explorer-in-Residence

NATIONAL GEOGRAPHIC **PHOTO GALLERY**

Regional photos, including Florence,
Paris, Amsterdam, and Budapest

Maps and Graphs
Interactive Map Tool

Interactive Whiteboard
GeoActivities
• Map Europe's Land Regions
• Compare Greek and Roman Governments
• Analyze Causes and Effects of World War I

Connect to NG
Research links and current events
in Europe

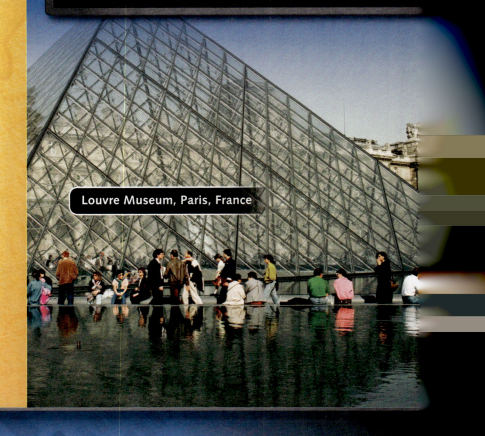

Louvre Museum, Paris, France

NATIONAL GEOGRAPHIC
ATLAS

**Page numbers of maps match Student Edition.
For a full Atlas go to the Teacher's Reference Guide.**

World Physical

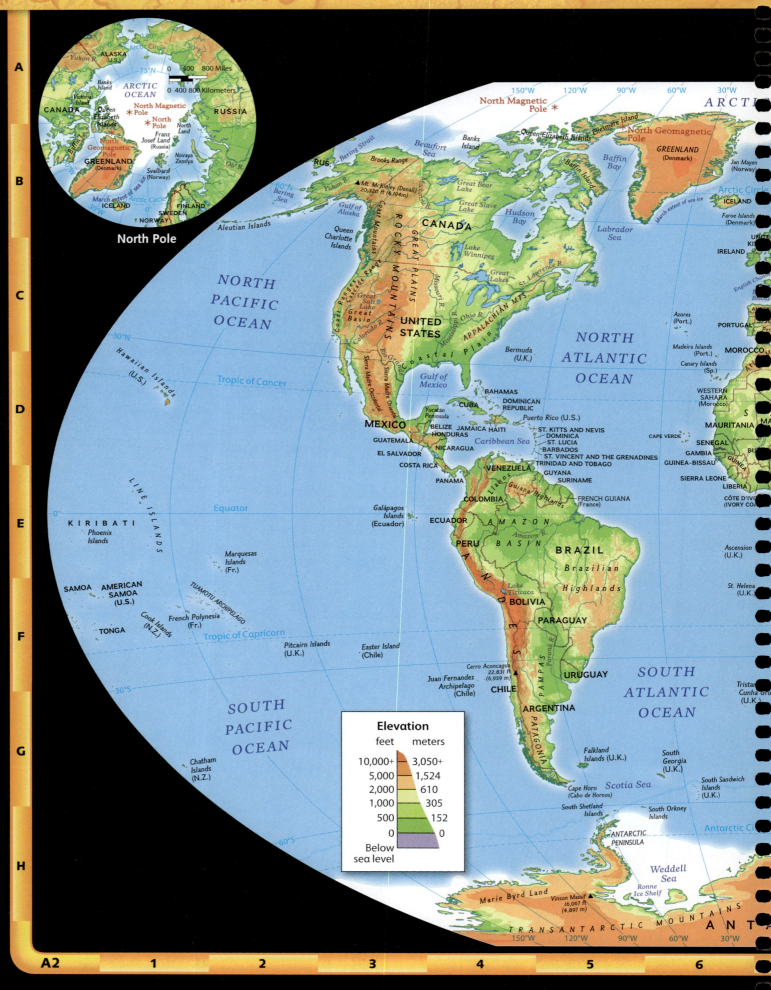

North Pole

Elevation

feet	meters
10,000+	3,050+
5,000	1,524
2,000	610
1,000	305
500	152
0	0
Below sea level	

South Pole

INDIAN OCEAN

NORTH PACIFIC OCEAN

ANTARCTICA

World Political

North Pole

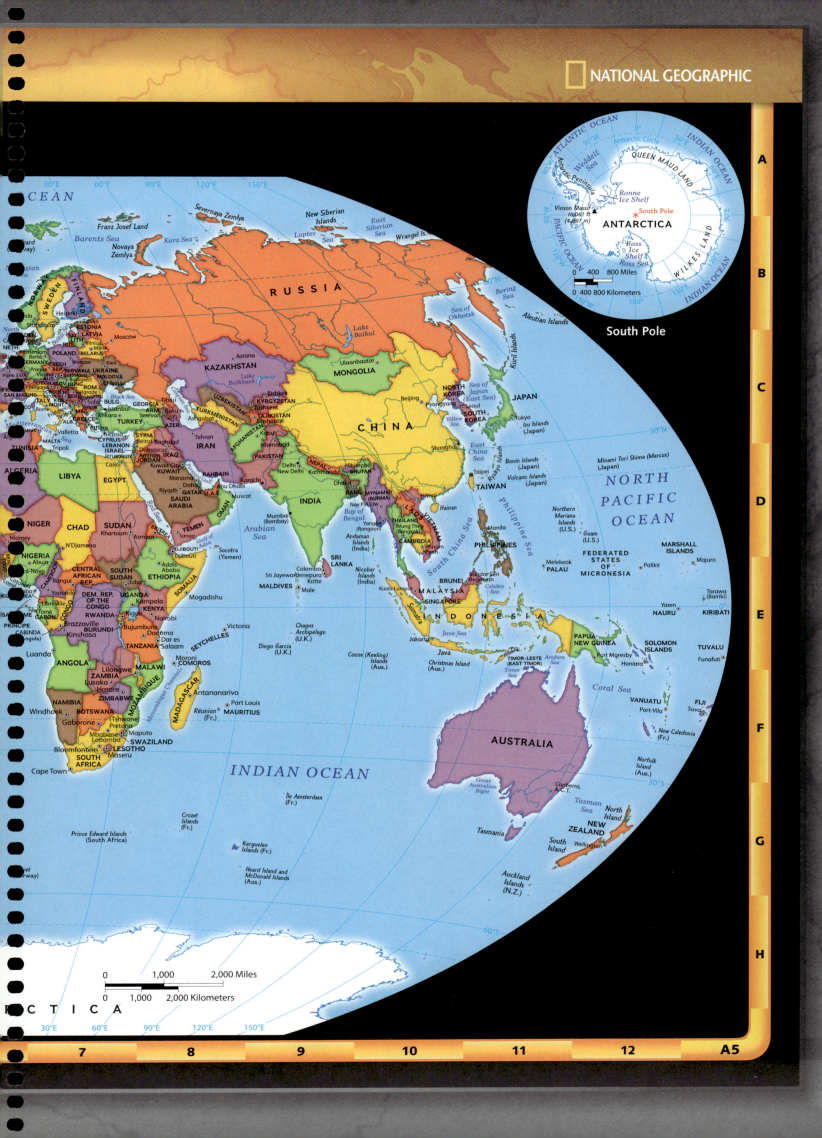

South Pole

ANTARCTICA

Ronne Ice Shelf

South Pole

Ross Ice Shelf

Ross Sea

QUEEN MAUD LAND

WILKES LAND

Antarctic Peninsula

Weddell Sea

Vinson Massif 16,067 ft (4,897 m)

0   400   800 Miles
0 400 800 Kilometers

RUSSIA

Franz Josef Land

Barents Sea

Novaya Zemlya

Kara Sea

Severnaya Zemlya

New Siberian Islands

Laptev Sea

East Siberian Sea

Wrangel Is.

NORWAY SWEDEN FINLAND

Moscow

ESTONIA LATVIA LITH. BELARUS

Bering Sea

Sea of Okhotsk

Aleutian Islands

KAZAKHSTAN

Astana

Lake Balkhash

MONGOLIA

Ulaanbaatar

Sea of Japan (East Sea)

Kuril Islands

POLAND UKRAINE MOLDOVA ROM. BULG.

GEORGIA ARM. AZER.

UZBEKISTAN

Bishkek Tashkent KYRGYZSTAN TAJIKISTAN

Beijing

NORTH KOREA Pyongyang

Seoul SOUTH KOREA

JAPAN Tokyo

Lake Baikal

TURKEY

SYRIA CYPRUS LEBANON ISRAEL JORDAN

TURKMENISTAN Ashgabat

AFGHANISTAN Kabul

CHINA

Yellow Sea Shanghai

Izu Islands (Japan)

Baghdad IRAQ KUWAIT

Tehran IRAN

Islamabad PAKISTAN

Delhi New Delhi

NEPAL Kathmandu BHUTAN Thimphu

East China Sea

Ryukyu Islands

TAIWAN Taipei

Bonin Islands (Japan)

Minami Tori Shima (Marcus) (Japan)

BAHRAIN QATAR U.A.E.

Riyadh SAUDI ARABIA

Abu Dhabi Muscat OMAN

BANG. Dhaka

MYANMAR (BURMA)

Hanoi

Hainan

Volcano Islands (Japan)

NORTH PACIFIC OCEAN

ALGERIA LIBYA EGYPT NIGER CHAD SUDAN

YEMEN Sanaa

Arabian Sea

Mumbai (Bombay)

INDIA

Bay of Bengal

Nay Pyi Taw

LAOS Vientiane

THAILAND Bangkok

VIETNAM CAMBODIA Phnom Penh

South China Sea

Northern Mariana Islands (U.S.)

Guam (U.S.)

FEDERATED STATES OF MICRONESIA

MARSHALL ISLANDS Majuro

NIGERIA Abuja

CENTRAL AFRICAN REP.

SOUTH SUDAN Juba

ETHIOPIA Addis Ababa

SOMALIA

DJIBOUTI

Socotra (Yemen)

Colombo Sri Jayewardenepura Kotte SRI LANKA

MALDIVES Male

Andaman Islands (India)

Nicobar Islands (India)

Yangon (Rangoon)

Manila

PHILIPPINES

Philippine Sea

Melekeok PALAU Palikir

Yaren NAURU

Tarawa (Bairiki) KIRIBATI

CAMEROON GABON DEM. REP. OF THE CONGO Kinshasa

UGANDA Kampala KENYA Nairobi RWANDA BURUNDI

TANZANIA Dodoma Dar es Salaam

SEYCHELLES

Victoria

Chagos Archipelago (U.K.)

Diego Garcia (U.K.)

Cocos (Keeling) Islands (Aus.)

Kuala Lumpur MALAYSIA BRUNEI Bandar Seri Begawan

SINGAPORE

Sumatra

Celebes Sea

Jakarta Java Sea

INDONESIA

Dili TIMOR-LESTE (EAST TIMOR)

PAPUA NEW GUINEA Port Moresby

SOLOMON ISLANDS Honiara

Yaren NAURU

TUVALU Funafuti

ANGOLA Luanda

ZAMBIA Lusaka

MALAWI Lilongwe

COMOROS Moroni

Antananarivo

MADAGASCAR

Réunion (Fr.) MAURITIUS Port Louis

Christmas Island (Aus.)

Java Arafura Sea Timor Sea

Coral Sea

VANUATU Port-Vila

FIJI Suva

New Caledonia (Fr.)

NAMIBIA Windhoek

BOTSWANA Gaborone

ZIMBABWE Harare

MOZAMBIQUE Maputo

Mbabane SWAZILAND Pretoria (Tshwane)

AUSTRALIA

Mozambique Channel

Lobamba Maseru LESOTHO

Bloemfontein SOUTH AFRICA

Cape Town

INDIAN OCEAN

Île Amsterdam (Fr.)

Crozet Islands (Fr.)

Prince Edward Islands (South Africa)

Kerguelen Islands (Fr.)

Heard Island and McDonald Islands (Aus.)

Great Australian Bight

Canberra, A.C.T.

Tasman Sea

Tasmania

North Island

NEW ZEALAND

South Island Wellington

Norfolk Island (Aus.)

Auckland Islands (N.Z.)

0   1,000   2,000 Miles
0   1,000   2,000 Kilometers

ARCTICA

Elevation

feet	meters
10,000+	3,050+
5,000	1,524
2,000	610
1,000	305
500	152
0	0
Below sea level	

ICELAND

Norwegian Sea

Arctic Circle

Faroe Islands (Denmark)

Shetland Islands (U.K.)

NORWAY

SCANDINAVIA

SWEDEN

Hebrides (U.K.)

Orkney Islands (U.K.)

NORTHERN IRELAND

SCOTLAND

North Sea

ATLANTIC OCEAN

IRELAND

Shannon R.

Irish Sea

UNITED KINGDOM

Gotland (Sweden)

JUTLAND

DENMARK

ZEALAND

Baltic

WALES

ENGLAND

NETHERLANDS

Elbe R.

NORTHERN

POLAND

Celtic Sea

Thames R.

English Channel

Channel Islands (U.K.)

BELGIUM

GERMANY

Rhine R.

Oder R.

LUXEMBOURG

Seine R.

Black Forest

Danube R.

CZECH REPUBLIC (CZECHIA)

Loire R.

FRANCE

SWITZERLAND

LIECHTENSTEIN

AUSTRIA

HUNG

Bay of Biscay

MASSIF CENTRAL

Mt. Blanc 15,781 ft.

(4,810 m)

Rhône R.

MONACO

Po R.

SLOVENIA

CROATIA

Cantabrian Mountains

PYRENEES

French Riviera

Ligurian Sea

SAN MARINO

BOSNIA AND HERZEGO

Douro R.

Iberian Mountains

Ebro R.

ANDORRA

Corsica (France)

ITALY

APENNINES

Adriatic Sea

MONTENEGRO

PORTUGAL

SPAIN

IBERIAN

Tagus R.

Guadiana R.

Balearic Sea

Sardinia (Italy)

VATICAN CITY

PENINSULA

Sierra Morena

Baetic Mountains

Balearic Islands (Spain)

Tyrrhenian Sea

Strait of Gibraltar

GIBRALTAR (U.K.)

Mediterranean

Sicily (Italy)

Ionian S

MOROCCO

ALGERIA

| 0 | 200 | 400 Miles |
| 0 | 200 | 400 Kilometers |

TUNISIA

MALTA

Sea

30°E 70°N 40°E 60°E 70°E 80°E

Barents Sea

KOLA PENINSULA

Pechora R.

Ob R.

Irtysh R.

White Sea

60°N

A

FINLAND

Northern Dvina R.

R U S S I A

B

Lake Onega

Gulf of Bothnia

70°E

Lake Ladoga

E U R O P E A N   P L A I N

Kama R.

C

Gulf of Finland

50°N

ESTONIA

Volga R.

N
W   E
S

KAZAKHSTAN

D

LATVIA

LITHUANIA

BELARUS

Ural R.

Aral Sea

60°E

E

AND

Vistula R.

Don R.

Volga R.

UZBEKISTAN

UKRAINE

Dnieper R.

Caspian Sea

F

Dniester R.

MOLDOVA

Sea of Azov

TURKMENISTAN

40°E

CARPATHIAN MOUNTAINS

CRIMEA

ARY

ROMANIA

Black Sea

GEORGIA

Danube R.

AZERBAIJAN

SERBIA

Balkan Mountains

ARMENIA

G

KOSOVO

BULGARIA

Bosporus

AZERB.

MACEDONIA

Sea of Marmara

NIA

Dardanelles

TURKEY

IRAN

GREECE

Aegean Sea

H

Euphrates R.

Tigris R.

Crete (Greece)

Rhodes (Greece)

CYPRUS

SYRIA

IRAQ

50°E 30°E

LEBANON

30°E 40°E

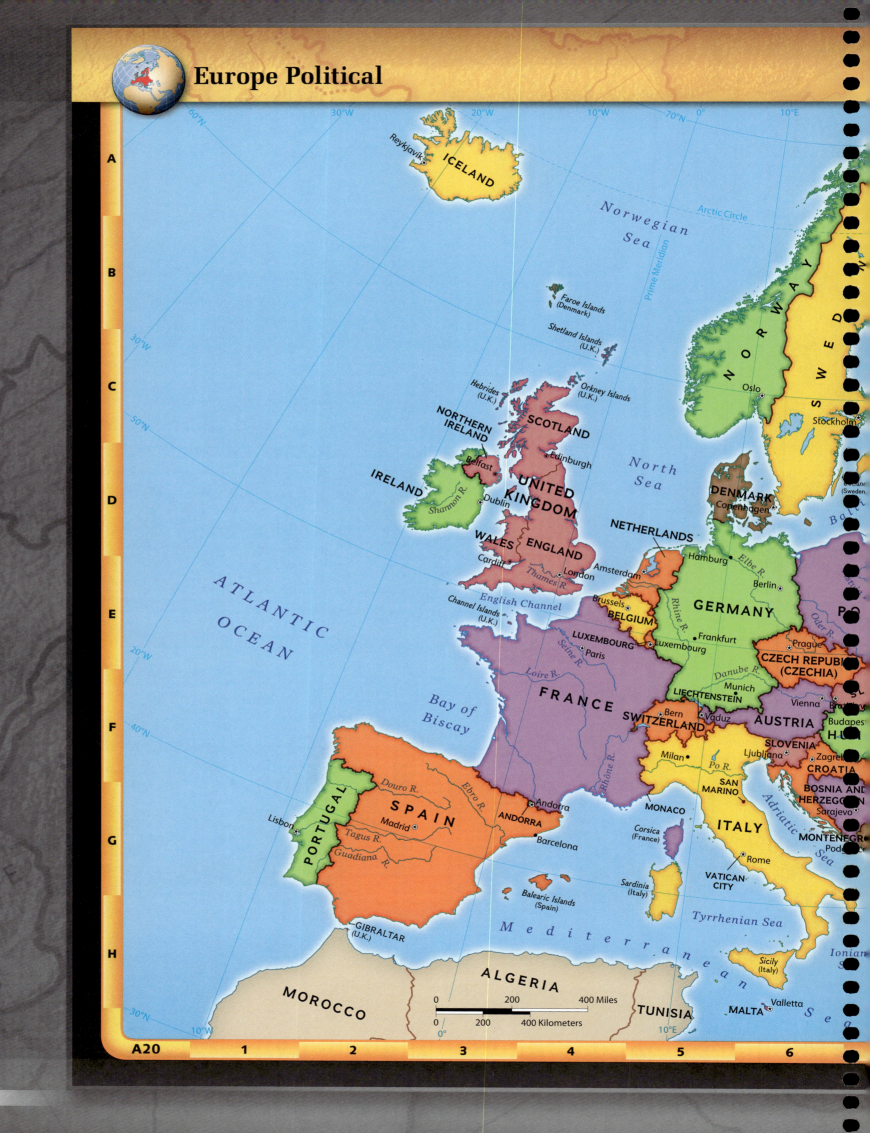

Europe Political

A B C D E F G H

A20 1 2 3 4 5 6

ICELAND
Reykjavik

Norwegian Sea

Arctic Circle

Faroe Islands (Denmark)

Shetland Islands (U.K.)

Hebrides (U.K.)
Orkney Islands (U.K.)

SCOTLAND
Edinburgh

NORTHERN IRELAND
Belfast

IRELAND
Shannon R.
Dublin

UNITED KINGDOM

WALES
Cardiff

ENGLAND
London
Thames R.

English Channel
Channel Islands (U.K.)

North Sea

NORWAY
Oslo

SWEDEN
Stockholm

DENMARK
Copenhagen

Rhine (Sweden)

NETHERLANDS
Amsterdam

Hamburg
Elbe R.
Berlin

GERMANY

BELGIUM
Brussels
Rhine R.

LUXEMBOURG
Luxembourg
Frankfurt

Paris
Seine R.

FRANCE
Loire R.

ATLANTIC OCEAN

Bay of Biscay

Rhône R.

Danube R.
Munich

CZECH REPUBLIC (CZECHIA)
Prague

LIECHTENSTEIN
SWITZERLAND
Bern
Vaduz

AUSTRIA
Vienna

SLOVENIA
Ljubljana

HUNGARY
Budapest

CROATIA
Zagreb

BOSNIA AND HERZEGOVINA
Sarajevo

MONTENEGRO
Podgorica

Adriatic Sea

Milan
Po R.

SAN MARINO

MONACO
ANDORRA
Andorra

ITALY

Corsica (France)

VATICAN CITY
Rome

PORTUGAL
Lisbon

Douro R.
Ebro R.

SPAIN
Madrid
Barcelona

Tagus R.
Guadiana R.

GIBRALTAR (U.K.)

Balearic Islands (Spain)

Sardinia (Italy)

Tyrrhenian Sea

Sicily (Italy)

MALTA
Valletta

Ionian Sea

Mediterranean Sea

MOROCCO

ALGERIA

TUNISIA

0   200   400 Miles
0   200   400 Kilometers

UNIT 2

EXPLORE EUROPE WITH NATIONAL GEOGRAPHIC

MEET THE EXPLORER

NATIONAL GEOGRAPHIC

Some archaeological sites are underwater. Emerging Explorer Katy Croff Bell works with archaeologists in the Mediterranean and Black seas to help them figure out where to look for submerged secrets.

INVESTIGATE GEOGRAPHY

The Alps are the highest and most extensive mountain range in Europe. They stretch across central Europe and are concentrated in France, Germany, Italy, Switzerland, and Austria. This is the Lauterbrunnen Valley in Oberland, Switzerland.

STEP INTO HISTORY

The Colosseum in Rome is one of the Roman Empire's greatest architectural and engineering achievements. The arena, completed in A.D. 80, seated nearly 50,000 spectators who watched gladiator games, performances, and even mock naval battles.

ONLINE WORLD ATLAS

London, United Kingdom

3,673 miles

Washington, D.C.

Go to **myNGconnect.com** for maps of Europe.

CONNECT WITH THE CULTURE

Architect I.M. Pei's modern pyramid serves as an entrance into the Louvre in Paris, France. The museum holds some of the world's greatest art treasures.

75

CHAPTER PLANNER

SECTION SUPPORT

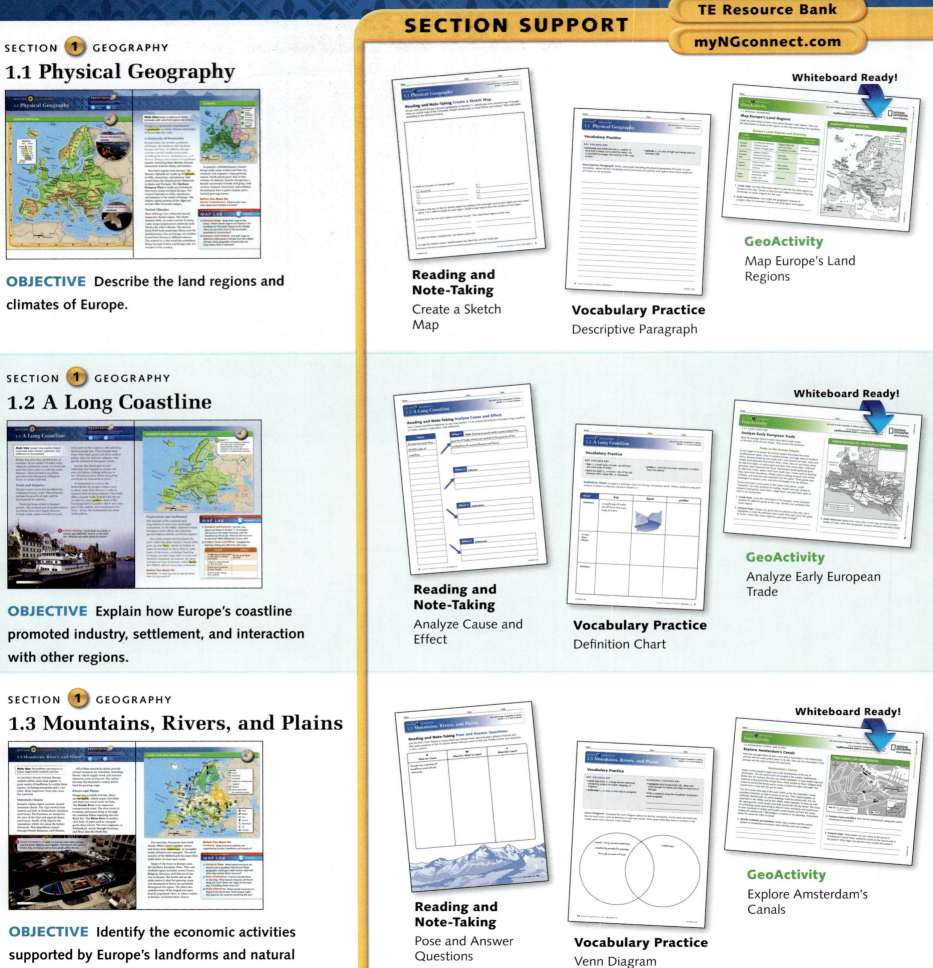

SECTION ① GEOGRAPHY
1.1 Physical Geography

OBJECTIVE Describe the land regions and climates of Europe.

Reading and Note-Taking
Create a Sketch Map

Vocabulary Practice
Descriptive Paragraph

Whiteboard Ready!

GeoActivity
Map Europe's Land Regions

SECTION ① GEOGRAPHY
1.2 A Long Coastline

OBJECTIVE Explain how Europe's coastline promoted industry, settlement, and interaction with other regions.

Reading and Note-Taking
Analyze Cause and Effect

Vocabulary Practice
Definition Chart

Whiteboard Ready!

GeoActivity
Analyze Early European Trade

SECTION ① GEOGRAPHY
1.3 Mountains, Rivers, and Plains

OBJECTIVE Identify the economic activities supported by Europe's landforms and natural resources.

Reading and Note-Taking
Pose and Answer Questions

Vocabulary Practice
Venn Diagram

Whiteboard Ready!

GeoActivity
Explore Amsterdam's Canals

ASSESSMENT

Student Edition
Ongoing Assessment: Map Lab

Resource Bank and myNGconnect.com
Review and Assessment, Sections 1.1–1.4

ExamView®
Test Generator CD-ROM
Section 1 Quiz in English and Spanish

Student Edition
Ongoing Assessment: Map Lab

Resource Bank and myNGconnect.com
Review and Assessment, Sections 1.1–1.4

ExamView®
Test Generator CD-ROM
Section 1 Quiz in English and Spanish

Student Edition
Ongoing Assessment: Map Lab

Resource Bank and myNGconnect.com
Review and Assessment, Sections 1.1–1.4

ExamView®
Test Generator CD-ROM
Section 1 Quiz in English and Spanish

TECHTREK  myNGconnect.com

 » Fast Forward!
Core Content Presentations
Teach *Physical Geography*

 Digital Library
NG Photo Gallery, Section 1

 Connect to NG
Research Links

Maps and Graphs
• **Interactive Map Tool**
Analyze Patterns of Population
• Online World Atlas: Europe Physical; Climate

Also Check Out
GeoJournal in
Student eEdition

 » Fast Forward!
Core Content Presentations
Teach *A Long Coastline*

 Digital Library
• GeoVideo: *Introduce Europe*
• NG Photo Gallery, Section 1

 Connect to NG
Research Links

Maps and Graphs
• **Interactive Map Tool**
Explore Mediterranean Trade
• Online World Atlas: Europe's Major Landforms and Rivers

Also Check Out
GeoJournal in
Student eEdition

 » Fast Forward!
Core Content Presentations
Teach *Mountains, Rivers, and Plains*

Maps and Graphs
Online World Atlas: Land Use and Natural Resources

 Digital Library
NG Photo Gallery, Section 1

Interactive Whiteboard
GeoActivity Explore Amsterdam's Canals

Also Check Out
• Graphic Organizers in
Teacher Resources
• GeoJournal in
Student eEdition

TE Resource Bank

myNGconnect.com

SECTION **1** GEOGRAPHY

1.4 Protecting the Mediterranean

OBJECTIVE Describe the human activities that are harming ecosystems in the Mediterranean Sea.

Whiteboard Ready!

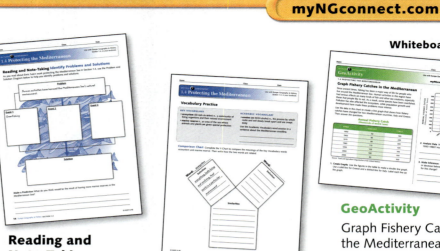

Reading and Note-Taking
Identify Problems and Solutions

Vocabulary Practice
Comparison Chart

GeoActivity
Graph Fishery Catches in the Mediterranean

SECTION **2** EARLY HISTORY

2.1 Roots of Democracy

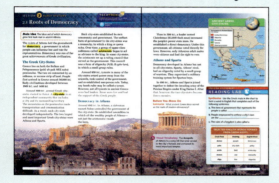

OBJECTIVE Identify and describe the influence of the ancient Greeks on the development of democracy.

Whiteboard Ready!

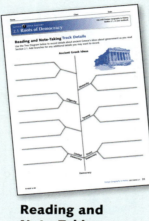

Reading and Note-Taking
Track Details

Vocabulary Practice
Definition and Details

GeoActivity
Analyze Primary Sources: Democracy

SECTION **2** EARLY HISTORY

2.2 Classical Greece

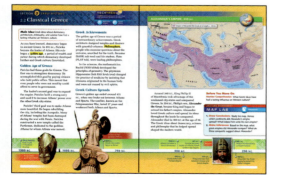

OBJECTIVE Explain how the cultural achievements of classical Greece influenced the ancient and modern worlds.

Whiteboard Ready!

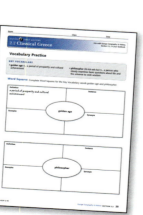

Reading and Note-Taking
Pose and Answer Questions

Vocabulary Practice
Word Squares

GeoActivity
Research Ancient Greek Contributions

ASSESSMENT

TECHTREK myNGconnect.com

Student Edition
Ongoing Assessment: Data Lab

Teacher's Edition
Performance Assessment: Conduct News Interviews

Resource Bank and myNGconnect.com
Review and Assessment, Sections 1.1–1.4

 ExamView®
Test Generator CD-ROM
Section 1 Quiz in English and Spanish

 Fast Forward!
Core Content Presentations
Teach *Protecting the Mediterranean*

 Digital Library
Explorer Video Clip: Enric Sala

 Connect to NG
Research Links

 Interactive Whiteboard
GeoActivity Graph Fishery Catches in the Mediterranean

Also Check Out
- Online World Atlas in **Maps and Graphs**
- Graphic Organizers in **Teacher Resources**
- GeoJournal in **Student eEdition**

Student Edition
Ongoing Assessment: Reading Lab

Resource Bank and myNGconnect.com
Review and Assessment, Sections 2.1–2.6

 ExamView®
Test Generator CD-ROM
Section 2 Quiz in English and Spanish

 Fast Forward!
Core Content Presentations
Teach *Roots of Democracy*

 Digital Library
NG Photo Gallery, Section 2

Maps and Graphs
Online World Atlas: Ancient Greek City-States

Interactive Whiteboard
GeoActivity Analyze Primary Sources: Democracy

Also Check Out
- Graphic Organizers in **Teacher Resources**
- GeoJournal in **Student eEdition**

Student Edition
Ongoing Assessment: Map Lab

Resource Bank and myNGconnect.com
Review and Assessment, Sections 2.1–2.6

 ExamView®
Test Generator CD-ROM
Section 2 Quiz in English and Spanish

 **Fast Forward!**
Core Content Presentations
Teach *Classical Greece*

 Digital Library
NG Photo Gallery, Section 2

 **Maps and Graphs**
Interactive Map Tool
Map Ancient Greece

Also Check Out
- Online World Atlas in **Maps and Graphs**
- Graphic Organizers in **Teacher Resources**
- GeoJournal in **Student eEdition**

CHAPTER PLANNER

SECTION 2 EARLY HISTORY

2.3 The Republic of Rome

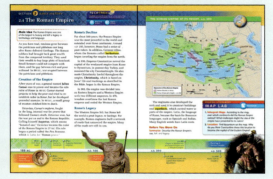

OBJECTIVE Compare the structure of the U.S. government with that of the Roman Republic.

Reading and Note-Taking
Find Main Idea and Details

Vocabulary Practice
Compare/Contrast Paragraph

Whiteboard Ready!

GeoActivity
Compare Greek and Roman Governments

SECTION 2 EARLY HISTORY

2.4 The Roman Empire

OBJECTIVE Analyze the rise and fall of the Roman Empire and the impact of Roman culture on Western civilization.

Reading and Note-Taking
Sequence Events

Vocabulary Practice
W-D-S Triangles

Whiteboard Ready!

GeoActivity
Analyze the Roots of Modern Languages

SECTION 2 EARLY HISTORY

2.5 Middle Ages and Christianity

OBJECTIVE Draw conclusions about life in the Middle Ages by analyzing the Roman Catholic Church and the feudal system.

Reading and Note-Taking
Summarize Information

Vocabulary Practice
Vocabulary Pyramid

Whiteboard Ready!

GeoActivity
Categorize Effects of the Crusades

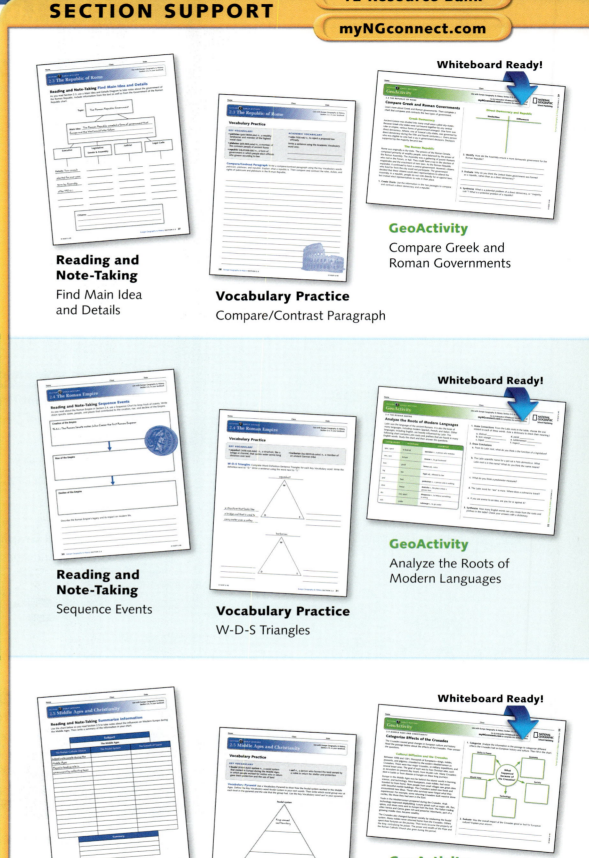

ASSESSMENT

Student Edition
Ongoing Assessment: Data Lab
Resource Bank and myNGconnect.com
Review and Assessment, Sections 2.1–2.6

ExamView®
Test Generator CD-ROM
Section 2 Quiz in English and Spanish

Student Edition
Ongoing Assessment: Map Lab
Resource Bank and myNGconnect.com
Review and Assessment, Sections 2.1–2.6

ExamView®
Test Generator CD-ROM
Section 2 Quiz in English and Spanish

Student Edition
Ongoing Assessment: Viewing Lab
Resource Bank and myNGconnect.com
Review and Assessment, Sections 2.1–2.6

ExamView®
Test Generator CD-ROM
Section 2 Quiz in English and Spanish

TECHTREK myNGconnect.com

  Fast Forward!
Core Content Presentations
Teach *The Republic of Rome*

Interactive Whiteboard
 GeoActivity Compare Greek
and Roman Governments

 Maps and Graphs
Online World Atlas: The Hills
of Rome

Digital Library
 NG Photo Gallery, Section 2

 Connect to NG
Research Links

Also Check Out
GeoJournal in
Student eEdition

 **Fast Forward!**
Core Content Presentations
Teach *The Roman Empire*

Interactive Whiteboard
GeoActivity Analyze the Roots
of Modern Languages

Connect to NG
Research Links

Also Check Out
• NG Photo Gallery in
  Digital Library
• Online World Atlas in
  Maps and Graphs
• GeoJournal in
  Student eEdition

  **Fast Forward!**
Core Content Presentations
Teach *Middle Ages and
Christianity*

Digital Library
NG Photo Gallery, Section 2

 Connect to NG
Research Links

Also Check Out
• Charts & Infographics and
  Graphic Organizers in
  Teacher Resources
• GeoJournal in
  Student eEdition

CHAPTER PLANNER

SECTION **2** EARLY HISTORY

2.6 Renaissance and Reformation

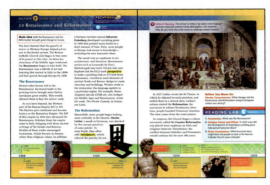

OBJECTIVE Analyze the cultural changes that took place in Europe during the Renaissance and Reformation.

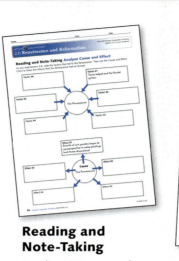

Reading and Note-Taking
Analyze Cause and Effect

Vocabulary Practice
Definition Chart

Whiteboard Ready!

GeoActivity
Map the Protestant Reformation

SECTION **3** EMERGING EUROPE

3.1 Exploration and Colonization

OBJECTIVE Describe European voyages of exploration and the impact of colonization.

Reading and Note-Taking
Outline and Take Notes

Vocabulary Practice
Words in Context

Whiteboard Ready!

GeoActivity
Compare European Explorers

SECTION **3** EMERGING EUROPE

3.2 The Industrial Revolution

OBJECTIVE Analyze and evaluate how industrialization changed European economies and people's way of life.

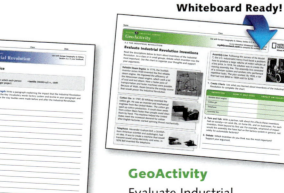

Reading and Note-Taking
Form and Support Opinions

Vocabulary Practice
Cause and Effect Paragraph

Whiteboard Ready!

GeoActivity
Evaluate Industrial Revolution Inventions

Student Edition
Ongoing Assessment: Reading Lab

Teacher's Edition
Performance Assessment: Quiz Each Other

Resource Bank and myNGconnect.com
Review and Assessment, Sections 2.1–2.6

 ExamView®
Test Generator CD-ROM
Section 2 Quiz in English and Spanish

 Fast Forward!
Core Content Presentations
Teach *Renaissance and Reformation*

 Digital Library
• GeoVideo: *Introduce Europe*
• NG Photo Gallery, Section 2

 Connect to NG
Research Links

Maps and Graphs
Interactive Map Tool
Analyze the Reformation's Legacy Today

Also Check Out
GeoJournal in
Student eEdition

Student Edition
Ongoing Assessment: Map Lab

Resource Bank and myNGconnect.com
Review and Assessment, Sections 3.1–3.6

 ExamView®
Test Generator CD-ROM
Section 3 Quiz in English and Spanish

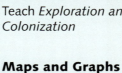 **Fast Forward!**
Core Content Presentations
Teach *Exploration and Colonization*

 Maps and Graphs
Online World Atlas: Early Colonization of Africa, Asia, and the Americas

Connect to NG
Research Links

Interactive Whiteboard
GeoActivity Compare European Explorers

Digital Library
NG Photo Gallery, Section 3

Also Check Out
GeoJournal in
Student eEdition

Student Edition
Ongoing Assessment: Map Lab

Resource Bank and myNGconnect.com
Review and Assessment, Sections 3.1–3.6

 ExamView®
Test Generator CD-ROM
Section 3 Quiz in English and Spanish

 Fast Forward!
Core Content Presentations
Teach *The Industrial Revolution*

 Digital Library
NG Photo Gallery, Section 3

 Connect to NG
Research Links

Maps and Graphs
Interactive Map Tool
Analyze the Impact of Industrialization

Also Check Out
• Online World Atlas in
Maps and Graphs
• GeoJournal in
Student eEdition

CHAPTER PLANNER

SECTION 3 EMERGING EUROPE

3.3 The French Revolution

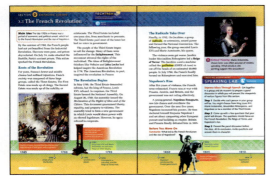

OBJECTIVE Summarize the causes and effects of the French Revolution and Napoleon's rise.

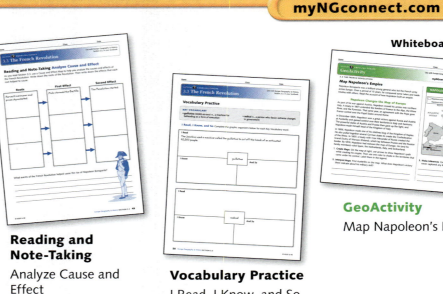

Whiteboard Ready!

Reading and Note-Taking
Analyze Cause and Effect

Vocabulary Practice
I Read, I Know, and So

GeoActivity
Map Napoleon's Empire

SECTION 3 DOCUMENT-BASED QUESTION

3.4 Declarations of Rights

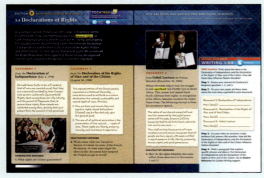

OBJECTIVE Analyze the philosophical ideas about human rights on which democracy is based.

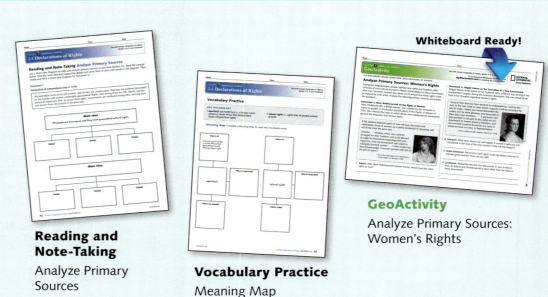

Whiteboard Ready!

Reading and Note-Taking
Analyze Primary Sources

Vocabulary Practice
Meaning Map

GeoActivity
Analyze Primary Sources: Women's Rights

SECTION 3 EMERGING EUROPE

3.5 Nationalism and World War I

OBJECTIVE Explain the nationalist tensions and struggles for power that led to World War I.

Whiteboard Ready!

Reading and Note-Taking
Analyze Cause and Effect

Vocabulary Practice
Definition Clues

GeoActivity
Analyze Causes and Effects of World War I

ASSESSMENT

Student Edition
Ongoing Assessment: Speaking Lab

Resource Bank and myNGconnect.com
Review and Assessment, Sections 3.1–3.6

ExamView®
Test Generator CD-ROM
Section 3 Quiz in English and Spanish

Student Edition
Ongoing Assessment: Writing Lab

Resource Bank and myNGconnect.com
Review and Assessment, Sections 3.1–3.6

ExamView®
Test Generator CD-ROM
Section 3 Quiz in English and Spanish

Student Edition
Ongoing Assessment: Map Lab

Resource Bank and myNGconnect.com
Review and Assessment, Sections 3.1–3.6

ExamView®
Test Generator CD-ROM
Section 3 Quiz in English and Spanish

TECHTREK myNGconnect.com

>> Fast Forward!
Core Content Presentations
Teach *The French Revolution*

Digital Library
NG Photo Gallery, Section 3

Also Check Out
• Graphic Organizers in **Teacher Resources**
• GeoJournal in **Student eEdition**

>> Fast Forward!
Core Content Presentations
Teach *Declarations of Rights*

Interactive Whiteboard
GeoActivity Analyze Primary Sources: Women's Rights

Teacher Resources
Graphic Organizers

Also Check Out
• NG Photo Gallery in **Digital Library**
• Writing Templates in **Teacher Resources**
• GeoJournal in **Student eEdition**

Connect to NG
Research Links

>> Fast Forward!
Core Content Presentations
Teach *Nationalism and World War I*

Connect to NG
Research Links

Digital Library
NG Photo Gallery, Section 3

Interactive Whiteboard
GeoActivity Analyze Causes and Effects of World War I

Maps and Graphs
Online World Atlas: Europe Before World War I, 1914

Also Check Out
• Graphic Organizers in **Teacher Resources**
• GeoJournal in **Student eEdition**

CHAPTER PLANNER

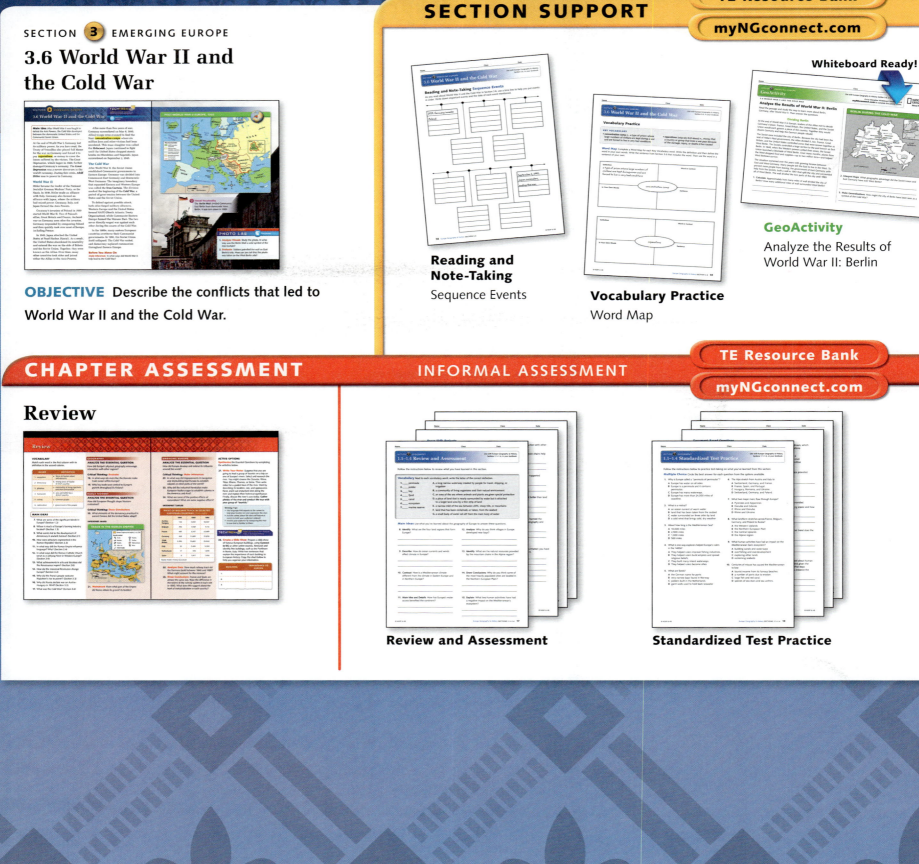

3.6 World War II and the Cold War

OBJECTIVE Describe the conflicts that led to World War II and the Cold War.

SECTION SUPPORT

TE Resource Bank

myNGconnect.com

Reading and Note-Taking
Sequence Events

Vocabulary Practice
Word Map

Whiteboard Ready!

GeoActivity
Analyze the Results of World War II: Berlin

CHAPTER ASSESSMENT

Review

INFORMAL ASSESSMENT

TE Resource Bank

myNGconnect.com

Review and Assessment

Standardized Test Practice

ASSESSMENT

Student Edition
Ongoing Assessment: Photo Lab

Teacher's Edition
Performance Assessment: Express Ideas
Through Speech

Resource Bank and myNGconnect.com
Review and Assessment, Sections 3.1–3.6

**ExamView®
Test Generator CD-ROM**
Section 3 Quiz in English and Spanish

TECHTREK myNGconnect.com

>> **Fast Forward!**

Core Content Presentations
Teach *World War II and the
Cold War*

Digital Library
GeoVideo: *Introduce Europe*

Connect to NG
Research Links

Maps and Graphs
Interactive Map Tool
Compare Europe Then and Now

Also Check Out
- NG Photo Gallery in **Digital Library**
- Online World Atlas in **Maps and Graphs**
- GeoJournal in **Student eEdition**

FORMAL ASSESSMENT

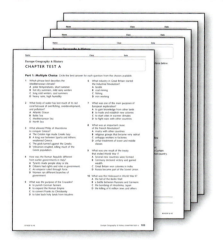

Chapter Test A (on level)

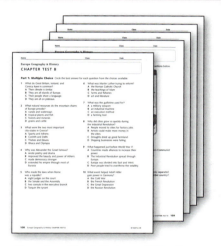

Chapter Test B (modified)

**ExamView®
Test Generator CD-ROM**
Chapter Tests

STRATEGIES FOR DIFFERENTIATION

Strategy 1 • Use a 3-2-1 Strategy

Display the 3-2-1 list below. Have students provide three of each item listed in the "3" column, two of each item listed in the "2" column, and one of each item listed in the "1" column.

3-2-1 Summary

Section	3	2	1
1.1	European peninsulas	major islands	climate region
1.2	effects of water access	bodies of water	use for polders
1.3	mountain chains	rivers	man-made water passage

Use with Sections 1.1–1.3 *You can also apply this activity to Sections 2.1–2.6. For example, in Section 2.1, you might ask students to provide three forms of government, two important city-states, and one achievement of Greek civilization.*

Strategy 2 • Play "Who Am I?"

Choose from the names below and distribute a list to students. Have them make game cards with a name on the front and a clue to the person's identity on the back. For example, for Socrates, students might write "leading ancient Greek philosopher." Use the cards to play a whole group, small group, or partner review game.

Julius Caesar	Charlemagne	Christopher Columbus
Jacques Cartier	Leonardo da Vinci	Pericles
Socrates	Sir Francis Drake	Johannes Gutenberg
Michelangelo	Plato	Aristotle
Adolf Hitler	John Locke	Alexander the Great
Louis XVI	Octavian	

Use with Sections 2.1–2.6 and 3.1–3.6

Strategy 3 • Use a K-W-L Chart

Before beginning each lesson in Section 2, have students work in pairs to preview the lesson title and complete these steps:

1. Write one fact they already know about the topic.
2. Write one question to which they want to know the answer.
3. After reading, answer the question they asked or discuss how they can find the answer. Then write one more fact they learned from reading that lesson.

Use with Sections 2.1–2.6

Strategy 4 • Build a Time Line

Select key events from Section 2.1. Then have students use the events to start a time line on the board. Students will add to the time line as they read Sections 2 and 3.

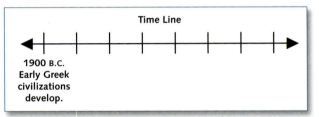

Use with Sections 2.1–2.6 and 3.1–3.6 *For example, key events from Section 2.1 might include Cleisthenes' establishment of direct democracy in Greece in 508 B.C. and the victory of Greece over the Persian Empire in 490 B.C.*

Strategy 5 • Use a Word Sort Activity

Write these words on the board and ask students to sort them into three groups of five related words each. Then have them use each group of words in a paragraph that shows how they are related.

exploration	guillotine	railroads	raw materials
inventions	injustice	revolution	child labor
unrest	colonies	third estate	
voyages	convert	factories	

Use with Sections 3.1–3.3

Strategy 1 • Modify Vocabulary Lists

Limit the number of vocabulary words, terms, and names students will be required to master. Have students write each word from your modified list on a colored sticky note and put it on the page next to where it appears in context.

Use with Sections 2.1–2.6

Strategy 2 • Sequence Events

Write events from Sections 2 and 3 on index cards. Read the events aloud and then have students put the cards in chronological order.

Use with Sections 2.1–2.6 and 3.1–3.6

ENGLISH LANGUAGE LEARNERS eEdition   Audiobook

Strategy **1** • Make Word Connections

Display the words below and ask students to talk about what each means and how they might be related. Then ask what the words might have to do with the upcoming chapter on Europe. Call on volunteers to use one of the words in a sentence about the location of Europe.

| rivers | seas | coastline | fjord |
| oceans | canals | bay | |

Use with Sections 1.1–1.3

Strategy **2** • Build Word Families

Remind students that words that have the same root and different suffixes belong to the same word families. Write the following word families on the board and have students volunteer to underline any suffixes that have been used to create related forms.

2.1 democracy, democrat, democratic
2.1 aristocracy, aristocrat, aristocratic

2.4 barbarian, barbarianism, barbaric
2.6 human, humanism, humanist, humanistic
3.5 nation, national, nationalism, nationalist, nationalistic

Use with Sections 2.1, 2.4, 2.6, and 3.5

Strategy **3** • Illustrate a Word Tree

Write the following word tree to help students understand the relationship of the groups. Then ask them to copy and draw pictures to illustrate each branch of the tree.

king
lord lord lord
vassal vassal vassal vassal
serf serf serf serf serf serf serf

Use with Section 2.5 *You might also have students write sentences explaining the relationships.*

GIFTED & TALENTED

Strategy **1** • Complete a Tic-Tac-Toe Project

Give students a choice of completing any three activities that form a tic-tac-toe win. Suggest that students develop a schedule for completing each part by an assigned end date.

Research and role-play a conversation between two knights of the Round Table.	Do a drawing of a medieval castle or cathedral. Label and explain the parts.	Do an oral report on the life of Alexander the Great. Tell why he does or does not deserve the title "great."
Create a display of works of art by famous Renaissance artists.	Read and retell a Greek or Roman myth.	Make a top-10 list of accomplishments of either Elizabeth I or Napoleon Bonaparte.
Read aloud the words of the French national anthem in English. Then play the music and give its history.	Do research and role-play with a partner an interview with a child working in an English textile factory.	Draw a map that shows the routes of three famous explorers from Europe to their final destinations.

Use with Sections 2.1–2.6

Strategy **2** • Read Historical Fiction

Work with the school librarian to create a display of historical fiction set in Europe during time periods such as classical Greece, the Roman Empire, the Middle Ages, the Renaissance, the French Revolution, World War I, or World War II. Allow students to choose and read a book and design a way to report on the book to the class.

Use with Sections 2.1–2.6 and 3.1–3.6 *As an alternative, you might ask groups of students to enact scenes from the books and bring particular historic events to life.*

PRE-AP Teacher Resources

Strategy **1** • Use the "Persia" Approach

Have students write an essay explaining the significance of the Middle Ages or the Renaissance. Copy the following mnemonic on the board, and tell students to use the "Persia" strategy to look at the period they choose:

Political—**E**conomic—**R**eligious—**S**ocial—**I**ntellectual—**A**rtistic

Use with Sections 2.5 and 2.6

Strategy **2** • Support an Opinion

Present a challenge to students to decide which explorer made the greatest impact on history. Have them develop a thesis statement that explains their decision and write an essay that supports it. Provide students with a writing template for the essay.

Use with Section 3.1

EUROPE
GEOGRAPHY & HISTORY

PREVIEW THE CHAPTER

Essential Question How did Europe's physical geography encourage interaction with other regions?

KEY VOCABULARY
- peninsula
- uplands
- polder
- bay
- fjord
- canal
- waterway
- ecosystem
- marine reserve

ACADEMIC VOCABULARY
navigable, erosion

TERMS & NAMES
- Northern European Plain
- Alps
- Danube River
- Rhine River

Essential Question How did European thought shape Western civilization?

KEY VOCABULARY
- democracy
- city-state
- golden age
- philosopher
- republic
- patrician
- plebeian
- barbarian
- aqueduct
- feudal system
- serf
- perspective
- indulgence

ACADEMIC VOCABULARY
aristocrat, veto

TERMS & NAMES
- Acropolis
- Alexander the Great
- Julius Caesar
- Augustus
- Christianity
- Middle Ages
- Crusades
- Renaissance
- Johannes Gutenberg
- Martin Luther
- Reformation
- Counter-Reformation

Essential Question How did Europe develop and extend its influence around the world?

KEY VOCABULARY
- navigation
- colony
- textile
- factory system
- radical
- guillotine
- natural rights
- apartheid
- nationalism
- trench
- reparations
- concentration camp

ACADEMIC VOCABULARY
convert, alliance

TERMS & NAMES
- Industrial Revolution
- Enlightenment
- John Locke
- Reign of Terror
- Napoleon Bonaparte
- Treaty of Versailles
- Great Depression
- Adolf Hitler
- Holocaust
- Iron Curtain
- Cold War
- Berlin Wall

76

TECHTREK
myNGconnect.com

 Fast Forward!
Core Content Presentations
Introduce *Europe Geography & History*

 **Digital Library**
- GeoVideo: *Introduce Europe*
- NG Photo Gallery

 **Maps and Graphs**
- **Interactive Map Tool**
  Analyze Geography and Movement
- Online World Atlas: Europe Political

Also Check Out
- Charts & Infographics in **Teacher Resources**
- Research Links in **Connect to NG**

INTRODUCE THE CHAPTER

INTRODUCE THE MAP

Use the Europe Political map to familiarize students with the countries in the region. Remind students of the difference between region and place. *(A place includes the characteristics of a specific location. A region is a group of places that have common characteristics.)*

Have students name countries on the map that they recognize. Ask them to point to and locate the countries on the map using the bingo index.

COMPARE ACROSS REGIONS

Download the Europe by the Numbers chart. **ASK:** Which European country on the chart has the highest population? *(Germany)* Which European country has the largest land area? *(Ukraine)* How do these countries compare to the United States? *(Both are many times smaller than the United States in both land and population.)* Encourage students to use the **Research Links** to discover more about the region.

EUROPE
BY THE NUMBERS

COUNTRY	LAND AREA (SQ MI)	POPULATION
France	212,345	62,814,233
Germany	134,623	82,282,988
Poland	117,474	38,463,689
Spain	192,657	46,505,963
Sweden	158,431	9,074,055
Ukraine	223,681	45,415,596
United Kingdom	93,410	62,348,447
United States	3,537,454	307,212,123

Source: CIA World Factbook

INTRODUCE THE ESSENTIAL QUESTIONS

SECTION 1 • GEOGRAPHY

How did Europe's physical geography encourage interaction with other regions?

Think, Pair, Share Activity: Geography and Movement This activity helps students predict the role geography has played in Europe's relationship with other regions. Give students this list of geographic terms: canyon, desert, mountain, ocean, plain, river. Tell students to think about ancient travelers living in a time before mechanized transport. Have them take a minute to consider the list of terms and decide which geographic features would aid travel between regions and which would act as barriers. Then have students form pairs and discuss the question. After five minutes, have each pair share their answers with the class. `0:15` minutes

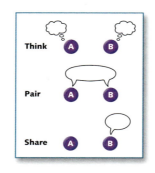

SECTION 2 • EARLY HISTORY

How did European thought shape Western civilization?

Four Corner Activity: Western Civilization This activity introduces students to four general aspects of Western civilization and allows them to choose which they think is the most influential. Post the four signs shown in the list below. Ask students to choose the aspect that they think has most shaped Western civilization, go to that corner, and then explain why. `0:20` minutes

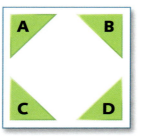

A Art European styles of painting, sculpture, music, and literature affected other cultures.

B Religion Europeans who migrated to other continents spread Christianity.

C Government Democratic and representative forms of government developed in Europe.

D Science European thinkers based science on observation and experiment, not religion.

SECTION 3 • EMERGING EUROPE

How did Europe develop and extend its influence around the world?

Roundtable Activity: International Influence This activity will allow students to explore the question by categorizing types of influence. Divide the class into groups of four or five students. Have the groups move desks together to form a table where they can all sit. Hand each group a sheet of paper with this question at the top: *What types of influence can one country have over another?* The first student in each group should write an answer and then pass the paper clockwise to the next student, who may add a new answer. The paper should be circulated around the group until the time is up. Students may pass at any time. After ten minutes, ask for volunteers to read their group's answers to the class. `0:20` minutes

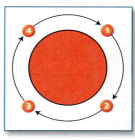

ACTIVE OPTIONS

Interactive Map Tool

Analyze Geography and Movement

PURPOSE Analyze how the features of Europe's physical geography affected human movement in ancient times

SET-UP

1. Open the **Interactive Map Tool,** set the "Region" to Europe and the "Map Mode" to Topographic.

2. Under "Physical Systems—Land," turn on the Surface Elevation layer. Set the transparency level to 50 percent.

ACTIVITY

ASK: Which pairs of countries have high mountains along their shared borders? *(France and Spain; Italy and France)* How would these mountains have affected movement between those countries? *(Mountains hindered movement.)* Then turn off the elevation layer. **ASK:** Why might it have been easier to travel from France to the United Kingdom than from France to Italy? *(It might have been easier to sail across the sea than to travel across the mountains.)* `0:15` minutes

INTRODUCE CHAPTER VOCABULARY

Word Sort Have students complete a Word Sort for this list of chapter vocabulary words:

aqueduct	factory system	reparations
bay	fjord	trench
canal	nationalism	uplands
city-state	peninsula	
democracy	polder	

Divide the class into groups and have them copy each word from the list on a separate note card or slip of paper. Then have the groups discuss what they think the words might mean and sort them into categories: **landform, body of water, human-made structure, economics, government**. Groups may also choose their own categories if they prefer. When groups have finished, ask them to explain their categories to the class.

SECTION **1** GEOGRAPHY

TECHTREK
myNGconnect.com For online maps
of Europe and Visual Vocabulary

Maps and
Graphs

Digital
Library

1.1 Physical Geography

EUROPE PHYSICAL

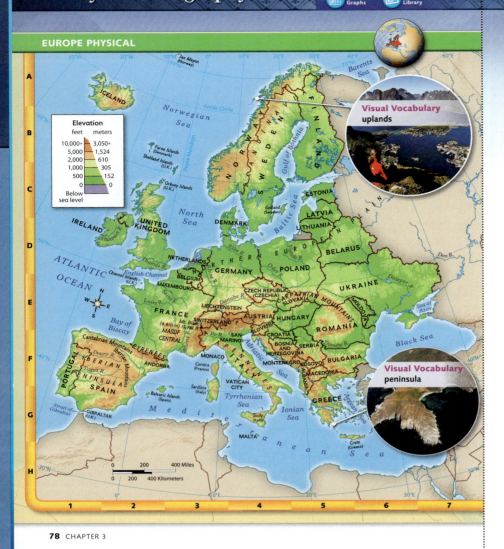

Elevation

feet	meters
10,000+	3,050+
5,000	1,524
2,000	610
1,000	305
500	152
0	0
Below sea level	

Visual Vocabulary
uplands

Visual Vocabulary
peninsula

Main Idea Europe is made up of several peninsulas with varied land regions and climates.

Europe is a "peninsula of peninsulas." A **peninsula** is a body of land surrounded on three sides by water.

A Peninsula of Peninsulas

Europe forms the western peninsula of Eurasia, the landmass that includes Europe and Asia. In addition, Europe contains several smaller peninsulas, including the Italian, Scandinavian, and Iberian. Europe also consists of significant islands, including Great Britain, Ireland, Greenland, Iceland, Sicily, and Corsica.

Four land regions form Europe. The Western Uplands are made up of **uplands**, or hills, mountains, and plateaus, that stretch from the Scandinavian Peninsula to Spain and Portugal. The **Northern European Plain** is made up of lowlands that reach across northern Europe. The Central Uplands are hills, mountains, and plateaus at the center of Europe. The Alpine region consists of the **Alps** and several other mountain ranges.

Varied Climates

Most of Europe lies within the humid temperate climate region. The North Atlantic Drift, an ocean current of warm water, keeps temperatures relatively mild. Winds also affect climate. The sirocco (shuh RAH koh) sometimes blows over the Mediterranean Sea and brings wet weather to southern Europe at different seasons. The mistral is a cold wind that sometimes blows through France and brings cold, dry weather to the country.

CLIMATE

Climate Regions
- Semiarid
- Humid Temperate– No dry season
- Dry Summer
- Humid Cold– No dry season
- Tundra & ice
- Unclassified highlands

In general, a Mediterranean climate brings mild, rainy winters and hot, dry summers and supports a long growing season. Hardy plants grow best in this climate. In contrast, Eastern Europe has a humid continental climate with long, cold winters. Iceland, Greenland, and northern Scandinavia have a polar climate and a limited growing season.

Before You Move On
Monitor Comprehension What are the main land regions and climates in Europe?

ONGOING ASSESSMENT
MAP LAB
GeoJournal

1. **Interpret Maps** Study both maps in this lesson. Which climate regions are found on the Scandinavian Peninsula? Based on the climate, where do you think most of the peninsula's population is concentrated?

2. **Compare and Contrast** Use both maps to determine what places in Europe have the coldest climates. What geographic characteristics do these places have in common?

land regions: Western Uplands, Central Uplands, Northern European Plain, Alpine region; climates: humid temperate, humid cold, semiarid, dry summer, tundra and ice

PLAN

OBJECTIVE Describe the land regions and climates of Europe.

CRITICAL THINKING SKILLS FOR SECTION 1.1

- Main Idea
- Monitor Comprehension
- Interpret Maps
- Compare and Contrast
- Analyze Visuals
- Explain
- Analyze Cause and Effect

PRINT RESOURCES

Teacher's Edition Resource Bank

- Reading and Note-Taking: Create a Sketch Map
- Vocabulary Practice: Descriptive Paragraph
- **GeoActivity** Map Europe's Land Regions

TECHTREK myNGconnect.com

Fast Forward!
Core Content Presentations
Teach *Physical Geography*

Digital Library
NG Photo Gallery, Section 1

Connect to NG
Research Links

Maps and Graphs
- **Interactive Map Tool**
Analyze Patterns of Population
- Online World Atlas: Europe Physical; Climate

Also Check Out
GeoJournal in
Student eEdition

BACKGROUND FOR THE TEACHER

Europe's physical geography facilitated movement into and out of the region. Unlike North America, Europe has no towering north-south mountain ranges that inhibited migration between east and west. As a result, in ancient times many groups moved from Asia into Europe and added to the cultural diversity of the continent. However, the migrations were not always peaceful. Throughout history, the plain that stretches across northern Europe acted as a wide open corridor for invading armies traveling in either direction.

ESSENTIAL QUESTION

How did Europe's physical geography encourage interaction with other regions?

Europe is a region of peninsulas, both large and small. Section 1.1 explains that some of these peninsulas form distinct regions with their own patterns of landforms and climate.

INTRODUCE & ENGAGE   Digital Library

Analyze Visuals Project the photos of the Danube River, the fjords in Norway, and an aerial shot of a European peninsula from the **NG Photo Gallery**. After showing all three photographs, **ASK:**

1. How are these photos of Europe similar to or different from the place where you live?
2. How do you think Europe's rivers and coastlines affect the way people earn a living?
3. In what ways do you think Europe's rivers and coastlines affect the way people travel? `0:15` minutes

TEACH   Maps and Graphs

Guided Discussion

1. **Explain** Why is Europe called a "peninsula of peninsulas"? (*Europe is a peninsula that extends westward from Asia. Smaller peninsulas also extend from the main peninsula.*)

2. **Analyze Cause and Effect** How do the bodies of water around Europe affect its climate? (*The North Atlantic Drift, a warm current in the Atlantic Ocean, helps keep temperatures warmer.*)

Interpret Maps Project the Europe Physical and Climate maps in the **Online World Atlas**. **ASK:** Which of Europe's four land regions is best suited for agriculture? Why? (*The Northern European Plain is best suited for agriculture because it is flat and easily farmed, and it has a humid temperate climate, which is good for many crops.*) `0:15` minutes

DIFFERENTIATE   Connect to NG

English Language Learners **Create Vocabulary Cards** Post the words *upland* and *lowland* where students can see them. Have students make vocabulary cards for these two words. On the front of each card, they should copy the word and draw a line between its two parts. On the back, they should create a simple drawing using arrows or other basic symbols to convey the word's meaning.

Pre-AP **Compare Peninsulas** Have students use the **Research Links** to research and create a chart that compares the Balkan, Crimean, Iberian, Italian, and Scandinavian peninsulas. Have them look for the following information:

- What bodies of water surround each peninsula?
- How large is each peninsula?
- What country or countries are located on each peninsula?
- What might be the advantages and disadvantages of living on a peninsula?

Have students design a chart like the one below to record the information.

	BALKAN	CRIMEAN	IBERIAN	ITALIAN	SCANDINAVIAN
Water					
Size					
Countries					

Interactive Map Tool

Analyze Patterns of Population

PURPOSE Analyze how physical geography influences where Europeans live

SET-UP

1. Open the **Interactive Map Tool**, set the "Region" to Europe and set the "Map Mode" to Terrain.

2. Under "Human Systems—Populations & Culture," turn on the Population Density layer. Set the transparency level to 40 percent.

ACTIVITY

1. Point out the map legend and make sure students understand what the shading on the map means. **ASK:** What landforms generally have a low population density? (*mountains*) On what landforms are many of the highest population densities found? (*coasts and plains*)

2. Under "Physical System—Climate," turn on the Climate Zones layer. Slide the transparency level to 70 percent, and then slide the transparency level for Population Density to 0 percent. **ASK:** What effect does climate have on population density? (*Fewer people live where climates are cold and harsh.*) `0:15` minutes

On Your Feet

Stand Up and Be Counted Have students get into groups of three. Give the groups a few minutes to study the physical map of Europe, looking for examples of peninsulas, islands, and mountain chains. Then have students number off within each group. Announce that 1s are mountains, 2s are peninsulas, and 3s are islands. Slowly read a list of geographic names that refer to specific landforms but do not give away their type. Examples include the Alps, Corsica, and Scandinavia. Students should stand each time they hear an example of their assigned landform and sit when other landforms are named. `0:15` minutes

ONGOING ASSESSMENT
MAP LAB   GeoJournal

ANSWERS

1. Climate regions include humid temperate, humid cold, and tundra and ice. People are most likely clustered in the humid temperate zones on the coasts.

2. The areas with the coldest climates are either close to or above the arctic circle or located in mountainous areas.

1.2 A Long Coastline

TECHTREK
myNGconnect.com For an online map and photos of Europe's coastal features and ports

Maps and Graphs   Digital Library

Main Idea Europe's long coastline helped to promote trade, industry, exploration, and settlement on the continent.

Europe has more than 24,000 miles of coastline. If you walked 25 miles a day along the continent's coasts, it would take more than four years to walk the entire distance. These extensive coastlines provided early Europeans with great access to oceans and seas.

Trade and Industry

Europe's water access has benefited the continent in many ways. These benefits include the growth of trade and the development of industry.

Trade has been central to Europe's growth. The civilizations of ancient Greece and Rome flourished largely because of trade. Early sailors traveled to nearly every port on the roughly 2,500-mile-long Mediterranean Sea. They brought back from other lands goods and ideas, such as grains, olive oil, and new religions, that greatly influenced European culture.

Europe also developed several industries that depend on oceans and seas, including a fishing industry. In fact, Europeans have fished along their coastlines for thousands of years.

In lowland areas such as the Netherlands, the people created a way to drain water from the sea in order to increase their farming industry. They built dikes, or giant walls, to hold back the sea in order to create **polders**. Most of the low-lying land of a polder, which once was part of the seabed, was transformed into farms. Today, the Netherlands has about 3,000 polders.

Critical Viewing A boat docks at a harbor in Gdansk (guh DANTSK), Poland, on the Baltic Sea. What do you notice about the harbor? The buildings are close to the edge of the water, and they look very old. The harbor is large and can accommodate big, modern ships.

EUROPE'S MAJOR LANDFORMS AND RIVERS

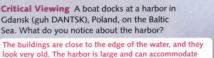

MAP TIP This map shows Europe's landforms and rivers, but it also includes country borders within the continent. You can use the map in Section 1.1 to identify the countries.

Exploration and Settlement

The location of the continent near large bodies of water also encouraged exploration. In the 1400s, explorers helped European rulers obtain raw materials, spread religious beliefs, and build empires.

Over time, people settled around the ports where the ships docked. Towns often grew up near **bays**, which are bodies of water surrounded on three sides by land. Some of the towns, including Hamburg, Germany, became large cities as trade and industry expanded. In contrast, the deep and narrow bays of Norway, called **fjords** (fee ORDZ), did not encourage settlement.

Before You Move On

Summarize In what ways has Europe benefited from its long coastline? The coastline has promoted trade, development of industries, growth of cities and towns, and exploration.

ONGOING ASSESSMENT
MAP LAB   GeoJournal

1. **Compare and Contrast** Use the map above and those in Section 1.1 to compare and contrast the Italian Peninsula with the Scandinavian Peninsula. What do the two have in common? What differences do you see?

2. **Analyze Cause and Effect** Complete the chart by writing one effect for each cause.

CAUSE	EFFECT
Trade was conducted on Europe's oceans and seas.	Goods and ideas spread.
Industry developed on the coasts.	
Explorers traveled to new lands.	
Cities grew along the coasts.	

PLAN

OBJECTIVE Explain how Europe's coastline promoted industry, settlement, and interaction with other regions.

CRITICAL THINKING SKILLS FOR SECTION 1.2

- Main Idea
- Summarize
- Compare and Contrast
- Analyze Cause and Effect
- Analyze Visuals
- Draw Conclusions
- Make Inferences
- Interpret Maps

PRINT RESOURCES

Teacher's Edition Resource Bank

- Reading and Note-Taking: Analyze Cause and Effect
- Vocabulary Practice: Definition Chart
- **GeoActivity** Analyze Early European Trade

TECHTREK   myNGconnect.com

Fast Forward!
Core Content Presentations
Teach *A Long Coastline*

Digital Library
- GeoVideo: *Introduce Europe*
- NG Photo Gallery, Section 1

Connect to NG
Research Links

Maps and Graphs
- **Interactive Map Tool** Explore Mediterranean Trade
- Online World Atlas: Europe's Major Landforms and Rivers

Also Check Out
GeoJournal in **Student eEdition**

BACKGROUND FOR THE TEACHER

The importance of trade in Europe dates back thousands of years, and some European ports are cities of great antiquity. Lisbon, for example, was probably founded about 1200 B.C. by Phoenician traders. Its original name, Olisipo, may have meant "delightful little port." Marseille, originally Massilia, was founded more than 2,500 years ago. Naples was originally a Greek colony. It was founded as Neopolis, or "new city," about 600 B.C.

ESSENTIAL QUESTION

How did Europe's physical geography encourage interaction with other regions?

Section 1.2 discusses how Europe's many peninsulas give it a great deal of coastline for its relatively small size, promoting sea travel and thus communication and trade.

INTRODUCE & ENGAGE   Digital Library

GeoVideo: *Introduce Europe* Show the portion of the video that presents Europe's coasts. **ASK:** In what ways might Europeans benefit from the continent's long coastline? *(enjoy its beauty; use it for recreation, fishing, and travel)*

Analyze Visuals Project the photos of coastlines in Gdansk and Amalfi from the NG Photo Gallery. **ASK:**

1. How are the buildings and landscapes of these two places similar and different?
2. What recreational activities can people do in a coastal city that they cannot do in an inland city? `0:10` minutes

TEACH   Maps and Graphs

Guided Discussion

1. **Draw Conclusions** From what other regions did Greek and Roman sailors bring back cultural influences to Europe? Explain. *(They brought back goods and ideas from around the Mediterranean, so the other regions that influenced Europe were probably North Africa and Southwest Asia.)*

2. **Make Inferences** Discuss the characteristics of a bay. **ASK:** Why would trading and fishing communities be more likely to be located on a bay than on a straight section of coast? *(Bays were bodies of water surrounded by land on three sides, so they provided more shelter for boats and ships to protect them from storms.)*

Interpret Maps Point out the Europe's Major Landforms and Rivers map in the text or the **Online World Atlas**. Suggest that students also use the physical map in Section 1.1 to identify countries. Have pairs of students work together to decide the shortest sailing routes between the following pairs of countries and to list the ports that ships would use. `0:15` minutes

United Kingdom to Norway:	*(Hartlepool)* →	*(Bergen)*
Spain to Italy:	*(Barcelona)* →	*(Genoa)*
Italy to France:	*(Genoa)* →	*(Marseille)*
France to United Kingdom:	*(Dunkirk)* →	*(London)*

DIFFERENTIATE   Connect to NG

Striving Readers **Analyze Cause and Effect** Remind students that an effect is the result of an action or condition. Allow striving readers to work in pairs with on-level readers to record the effects of Europe's long coastline on a Cause-and-Effect Web. Point out that the blue headings in the lesson will provide them with important clues about types of effects to look for.

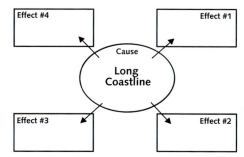

Gifted & Talented **Draw a Polder** Have students use the **Research Links** to learn about the process of making a polder. Suggest that they take notes about the steps the Dutch follow to reclaim land from the sea. Then they should make an illustrated process diagram by creating a drawing for each important step and arranging them in sequence.

ACTIVE OPTIONS

Interactive Map Tool

Explore Mediterranean Trade

PURPOSE Explore the distances sailors traveled to trade around the Mediterranean

SET-UP

1. Open the **Interactive Map Tool**, set the "Region" to Europe, and set the "Map Mode" to Topographic.
2. Zoom in and center the map on the Mediterranean region.
3. Turn on the Measure tool by clicking on the ruler icon at the upper right.

ACTIVITY

1. Click once on Athens, and then double click on Alexandria. Ask a volunteer to read out the distance between the two ports.
2. Repeat the process by measuring the distance between Naples and Palermo; Genoa and Algiers; Marseille and Barcelona.
3. **ASK:** What might be an advantage of trade between European ports? *(Possible response: The distances are usually shorter, so travel was probably safer.)* What might be an advantage of trade between Europe and Africa? *(Possible response: It is more likely traders could find a wider variety of goods by sailing to another continent.)* `0:20` minutes

On Your Feet

Inside-Outside Circle Have students stand in concentric circles facing each other. Have students in the outside circle ask students in the inside circle a question about the lesson. Then have the outside circle rotate one position to the right to create new pairings. After five questions, have students switch roles and continue. `0:15` minutes

---

ONGOING ASSESSMENT

MAP LAB   GeoJournal

ANSWERS

1. Both are surrounded on three sides by water, but the Scandinavian Peninsula is much bigger than the Italian Peninsula.

2.

CAUSE	EFFECT
Trade was conducted on Europe's oceans and seas.	Goods and ideas spread.
Industry developed on the coasts.	Fishing, shipbuilding, and farm industries developed.
Explorers traveled to new lands.	Europe extended its power.
Cities grew along the coasts.	Cities became centers of trade.

1.3 Mountains, Rivers, and Plains

Main Idea The landforms and resources in Europe support many economic activities.

As you have already learned, Europe consists of four main land regions. A great variety of landforms lie within these regions, including mountains and a vast plain. Many important rivers also cross the continent.

Mountain Chains

Europe's Alpine region contains several mountain chains. The Alps stretch from Austria and Italy to Switzerland, Germany, and France. The Pyrenees are located to the west of the Alps and separate Spain and France. South of the Alps lie the Apennines, which run along the Italian Peninsula. The Carpathians extend through Poland, Romania, and Ukraine.

All of these mountain chains provide natural resources for industries, including forests, which supply wood, and mineral resources, such as iron ore. The valleys between the mountains contain fertile land for growing crops.

Rivers and Plains

Europe has a wealth of rivers. Many are **navigable**, which means that boats and ships can travel easily on them. The **Danube River** is an important transportation route. The river starts in Germany and passes along or through ten countries before emptying into the Black Sea. The **Rhine River** is another vital body of water used to transport goods deep inland. The river originates in Switzerland, winds through Germany, and flows into the North Sea.

▼ **Visual Vocabulary** A **canal** is a human-made water passage used for travel, shipping, and irrigation. The boats in this canal in Venice, Italy, are being used to move goods within the city.

LAND USE AND NATURAL RESOURCES

Land Use
- Forest
- Woodland
- Grassland
- Mixed-use, including crops
- Cropland
- Wetland
- Desert, barren land
- Ice, cold desert, tundra

Natural Resources
- Coal
- Copper
- Fish
- Iron ore
- Natural gas
- Oil
- Uranium

For centuries, Europeans have built canals. When linked together, canals and rivers form **waterways**, or navigable routes of travel and transport. The small country of the Netherlands has more than 3,000 miles of rivers and canals.

Many of the rivers in Europe cross the Northern European Plain. This vast lowland region stretches across France, Belgium, Germany, and Poland all the way to Russia. The fertile soil on the plain makes it ideal for growing crops, and thousands of farms are sprinkled throughout the region. The plain also contains some of the largest and most heavily populated cities, or urban centers, in Europe, including Paris, France.

Landforms support trade, transportation, and agriculture. Natural resources supply wood and iron ore.

Before You Move On

Summarize What economic activities are supported by Europe's landforms and resources?

ONGOING ASSESSMENT

MAP LAB   GeoJournal

1. **Interpret Maps** What natural resources are found in the Carpathian Mountains? What geographic challenges might workers deal with when they extract these resources?

2. **Draw Conclusions** Find the Danube River on the map. What natural resources are found along the river? What role might the Danube play in handling these resources?

3. **Make Inferences** What natural resources are found in the North Sea? What impact might they have on the countries bordering the sea?

PLAN

OBJECTIVE Identify the economic activities supported by Europe's landforms and natural resources.

CRITICAL THINKING SKILLS FOR SECTION 1.3

- Main Idea
- Summarize
- Interpret Maps
- Draw Conclusions
- Make Inferences
- Synthesize

PRINT RESOURCES

Teacher's Edition Resource Bank

- Reading and Note-Taking: Pose and Answer Questions
- Vocabulary Practice: Venn Diagram
- **GeoActivity** Explore Amsterdam's Canals

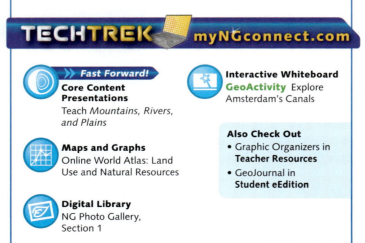

TECHTREK myNGconnect.com

▶▶ **Fast Forward!**
Core Content Presentations
Teach *Mountains, Rivers, and Plains*

Maps and Graphs
Online World Atlas: Land Use and Natural Resources

Digital Library
NG Photo Gallery, Section 1

Interactive Whiteboard
GeoActivity Explore Amsterdam's Canals

Also Check Out
- Graphic Organizers in **Teacher Resources**
- GeoJournal in **Student eEdition**

BACKGROUND FOR THE TEACHER

The Alps, Europe's highest mountain range, are not just an empty region of rock, snow, and ice. The region has a history of farming in its high valleys, but that economic activity has declined in recent years. Since the 1800s, industries such as steel mills and paper plants have grown more important. Also, since 1960, the major economic development in the Alps has been the growth of mass tourism. People visit the Alps by the thousands.

ESSENTIAL QUESTION

How did Europe's physical geography encourage interaction with other regions?

Rivers provide the means to transport the products that result from the fertile land and natural resources to the coast and other markets. Section 1.3 discusses transportation and economic activities in Europe.

INTRODUCE & ENGAGE

Activate Prior Knowledge Explain that in this section, students will learn about important geographic features of Europe. Have them use their geographic knowledge of the United States to answer the questions below. Encourage students to use a chart like the one shown here to organize their ideas.

Mountains	Rivers

- What are the major mountain chains of the United States?
- What types of activities are supported by these mountains?
- What are some major rivers of the United States?
- How do people use those rivers? `0:10` minutes

TEACH — Maps and Graphs

Guided Discussion

1. **Make Inferences** Are products shipped between France and Spain more likely to go overland or by sea? Explain. *(They are more likely to go by sea because the Pyrenees act as a land barrier between the two countries.)*

2. **Synthesize** In what ways do Europe's canals contribute to the continent's economy? *(The canals help the economy because they make it easier to ship goods where there are no rivers.)*

3. **Interpret Maps** Project the Land Use and Natural Resources map from the **Online World Atlas**. **ASK:** Which landform accounts for much of the large area of cropland on the map? *(the Northern European Plain)*

MORE INFORMATION

Rivers and Mountains Many of the rivers of Europe originate in uplands and mountains and flow down to the surrounding seas. The Alps, Apennines, and Carpathians act as watersheds, meaning that rivers on either side of the mountain chain drain in opposite directions. In spring, melting snow in the mountains flows into the rivers increasing their volume. This process helps replenish the rivers in the lowlands when demand for their use is highest.

The Rhine River

DIFFERENTIATE

Inclusion **Locate Mountains and Rivers** Give pairs of students this list of terms: Alps, Apennines, Carpathian, Caucasus, Danube, Pyrenees, Rhine. Then help the pairs find the terms on the map. Help them describe the location of these major landforms and rivers.

English Language Learners **Explain Multiple-Meaning Words** Alert students to the presence of two multiple-meaning words in this lesson. Point out the phrase *mountain chain*. Students may be familiar with the definition of *chain* as "a length of linked metal loops." However, in the lesson, a chain refers to a series of connected mountains. Also explain that in this lesson, the verb *stretch* does not mean "to pull something to make it longer." Instead, it means "to extend."

ACTIVE OPTIONS

Interactive Whiteboard
GeoActivity

Explore Amsterdam's Canals Have students work in pairs to answer the map questions and critical thinking questions to understand how canals have affected life in Amsterdam. Allow pairs to compare their answers with those of another group. Then each pair should write a paragraph that a tour guide could use to explain the importance of Amsterdam's canals to visitors. Ask volunteers to read their paragraphs to the rest of the class. `0:20` minutes

NG Photo Gallery

Analyze Visuals Have students view photos of the Alps, Danube, and canals in Venice from the **NG Photo Gallery**. Invite students to share their impressions of these landforms and waterways. **ASK:** Why do you think so many people visit the Alps each year? *(to enjoy the scenery, ski, swim, hike)* Based on the photos, why is the Danube such an important river in Europe? *(because it's wide enough for large ships; because it flows through major cities)* `0:15` minutes

On Your Feet

Three-Step Interview Have students choose a partner. One student should interview the other on the question *How does Europe's physical geography affect its economy?* Then they should reverse roles. Finally, each student should share the results of his or her interview with the class. `0:15` minutes

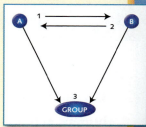

ONGOING ASSESSMENT
MAP LAB — GeoJournal

ANSWERS

1. forest products, coal, natural gas, and oil; moving resources out of the mountains might be difficult
2. coal, natural gas, and uranium; goods can be transported along the Danube to the Black Sea
3. fish, oil, and natural gas; these resources can help supply the surrounding countries with food and energy

1.4

SECTION 1 GEOGRAPHY
NATIONAL GEOGRAPHIC

TECHTREK
myNGconnect.com For a map, photos, and an Explorer Video Clip

Maps and Graphs   Digital Library

Protecting the Mediterranean
with Enric Sala

Main Idea Human activities have harmed the Mediterranean Sea's natural environment.

Under the Sea

On June 4, 2010, National Geographic Fellow Enric Sala began an exciting expedition: exploring the underwater world of the Mediterranean Sea. He wanted to find out how human activities have affected the sea's ecosystem.

An **ecosystem** is a community of living organisms and their natural environment. Three human activities have had an impact on the Mediterranean's ecosystem. One is overfishing, which occurs when people catch fish at a faster rate than the fish can reproduce. Another is pollution. The third is overdevelopment, which has occurred as coastal populations have grown. (See the chart on the next page.) Growing populations have added to the Mediterranean's pollution and the **erosion**, or wearing away, of its coastline.

myNGconnect.com
For more on Enric Sala in the field today

Critical Viewing Sala swims with a sea turtle. What equipment is he using during this underwater exploration?

Sala is wearing scuba gear, with an oxygen tank and goggles. He also appears to be equipped with a video camera and microphone.

MEDITERRANEAN EUROPE

Sala's work was inspired by Jacques Cousteau, the French underwater explorer. When he became an explorer himself, Sala sailed with Cousteau's son, Pierre-Yves Cousteau. They compared the condition of the Mediterranean Sea now with its condition 65 years earlier and found that the sea has been damaged. Sala concluded, "We have lost most of the large fish and the red coral because of centuries of exploitation [misuse]."

Marine Reserves

In spite of the harm that has been done, Sala sees signs of hope. During their Mediterranean expedition, he and Cousteau visited the Scandola Natural Reserve, near Italy. Scandola is a **marine reserve**, or protected area where people are prohibited from fishing, swimming, or anchoring their boats. As a result, marine life is thriving at the reserve. "This marine reserve," Sala said, "has restored the richness that Jacques Cousteau showed us 65 years ago."

Before You Move On

Monitor Comprehension What can be done to protect the Mediterranean Sea from the human activities that have harmed its environment?

take steps to reduce pollution, overfishing, and overdevelopment

POPULATION OF MEDITERRANEAN CITIES (IN MILLIONS)		
City	1960	2015 (projected)
Athens, Greece	2.2	3.1
Barcelona, Spain	1.9	2.73
Istanbul, Turkey	1.74	11.72
Marseille, France	0.8	1.36
Rome, Italy	2.33	2.65

Source: UN, 2002

ONGOING ASSESSMENT
DATA LAB
GeoJournal

1. **Interpret Charts** Based on the chart, which Mediterranean city will have undergone the greatest growth by 2015? What impact will this growth have on the city's coastline?

2. **Make Inferences** According to the chart, which city will have undergone the least growth by 2015? What does this projected number mean for the city's coastline?

3. **Turn and Talk** What could be done today to preserve fish populations for the future? Get together with a partner and come up with one or two specific suggestions. Share your ideas with the rest of the class.

SECTION 1.4 **85**

PLAN

OBJECTIVE Describe the human activities that are harming ecosystems in the Mediterranean Sea.

CRITICAL THINKING SKILLS FOR SECTION 1.4

- Main Idea
- Monitor Comprehension
- Interpret Charts
- Make Inferences
- Pose Questions
- Draw Conclusions
- Identify Solutions
- Create Graphs

PRINT RESOURCES

Teacher's Edition Resource Bank

- Reading and Note-Taking: Identify Problems and Solutions
- Vocabulary Practice: Comparison Chart
- **GeoActivity** Graph Fishery Catches in the Mediterranean

TECHTREK myNGconnect.com

Fast Forward!
Core Content Presentations
Teach *Protecting the Mediterranean*

Digital Library
Explorer Video Clip: Enric Sala

Connect to NG
Research Links

Interactive Whiteboard
GeoActivity Graph Fishery Catches in the Mediterranean

Also Check Out
- Online World Atlas in **Maps and Graphs**
- Graphic Organizers in **Teacher Resources**
- GeoJournal in **Student eEdition**

BACKGROUND FOR THE TEACHER

One Mediterranean species that has declined in numbers is red coral, a marine animal that lives in colonies, particularly in shallow waters. Since ancient times, people have harvested red coral from the sea so that they could use the skeleton in jewelry. The species is very slow growing, so it has difficulty recovering when it is overharvested. Other human activities, such as nearby marine construction, have also been shown to damage coral colonies.

ESSENTIAL QUESTION

How did Europe's physical geography encourage interaction with other regions?

Section 1.4 explores how European countries all contribute to the degradation of the Mediterranean Sea as a result of human activities.

INTRODUCE & ENGAGE 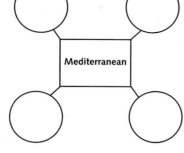 Digital Library

Explorer Video Clip: *Enric Sala* Show the **Explorer Video Clip** to provide background information on Enric Sala and his work in the Mediterranean. Ask students to summarize Sala's work and what he hopes to accomplish.

Pose Questions Have students use an Idea Web like the one shown to jot down questions about the Mediterranean Sea that they think the lesson will answer. If needed, start them off with an example, such as "Why does the Mediterranean need protecting?" `0:15` **minutes**

TEACH

Guided Discussion

1. **Draw Conclusions** Have volunteers find each city listed in the chart on the Mediterranean Europe map. Discuss the impact of the increasing population in each city. **ASK:** How are overdevelopment and pollution related? *(As more people live in a place, they produce more waste. They also build more industries, which may cause increased pollution.)*

2. **Identify Solutions** According to Sala, what is one thing people can do to help restore the Mediterranean? *(set aside marine reserves)*

Create Graphs Have pairs of students turn the data in the chart into a bar graph. Suggest that they choose two colors, one to represent population in 1960 and one to represent population in 2015. Each city will have two bars, one of each color. After students have finished, **ASK:** How does the graph help you see the amount each city grew? *(The length of the bars represents the amount of growth.)* `0:20` **minutes**

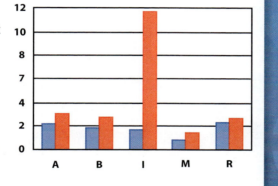

DIFFERENTIATE  Connect to NG

Striving Readers Find Details Have students copy the main idea from the lesson into the top box of a graphic organizer like the one below. Then have them scan the text to find and record three types of human activities that have affected the Mediterranean Sea's environment.

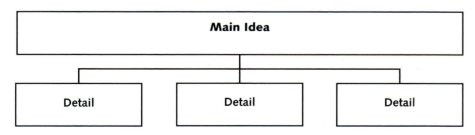

Gifted & Talented Stop Erosion Have students work in small groups and use the **Research Links** to research methods of preventing coastal erosion. Then ask them to write and act out a public service announcement about how to save the Mediterranean coast.

 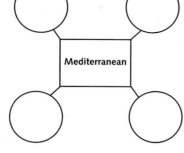

NATIONAL GEOGRAPHIC
Photo Gallery • Scotland's Rugged Coast

For more photos from the National Geographic Photo Gallery, go to the **Digital Library** at myNGconnect.com.

The Carnival in Venice

Ancient Roman aqueduct

Musicians in Krakow, Poland

Critical Viewing These rocky pinnacles, or pointed formations, took shape thousands of years ago on a hill in Scotland called the Storr. The largest of the formations shown here is known as the Old Man of Storr.

Italy's Amalfi Coast

French pastries on display

The Parthenon in Athens

Armor from the 1600s

2.1 Roots of Democracy

TECHTREK
myNGconnect.com For an online map of ancient Greek city-states

Maps and Graphs

Main Idea The ideas out of which democracy grew first took root in ancient Athens.

The rulers of Athens laid the groundwork for **democracy**, a government in which people can influence law and vote for representatives. Democracy was one of the great achievements of Greek civilization.

The Greek City-States

Greece lies on both the Balkan and Peloponnesus (pehl uh puh NEE suhs) peninsulas. The two are connected by an isthmus, or narrow strip of land. People first arrived in Greece around 50,000 B.C. Early civilizations developed between 1900 B.C. and 1400 B.C.

Around 800 B.C., several Greek city-states started to thrive. A **city-state** is an independent community that includes a city and its surrounding territory. The mountains on the peninsulas made transportation and communication difficult. As a result, each city-state developed independently. The two largest and most important Greek city-states were Athens and Sparta.

Each city-state established its own community and government. The earliest form of government in the city-states was a monarchy, in which a king or queen rules. Over time, a group of upper-class noblemen called **aristocrats** began to act as advisors to the king. In some city-states, the aristocrats set up a ruling council that served as the government. This council was a form of oligarchy (AHL ih gahr kee), in which a small group rules.

Around 650 B.C., tyrants in many of the city-states seized power away from the councils, took control of the government, and re-established one-person rule. Today, any harsh ruler may be called a tyrant. However, not all tyrants in ancient Greece were bad leaders. Some were fair and had the support of the Greek people.

Democracy in Athens

Around 600 B.C. in Athens, a statesman named Solon controlled the government of the city-state. He established assemblies in which all the wealthy people of Athens—not just the aristocrats—made the laws.

Then in 508 B.C., a leader named Cleisthenes (KLIHS thuh neez) increased the people's power even more. He established a direct democracy. Under this government, all citizens voted directly for laws. However, only Athenian adult males were citizens and had the right to vote.

Athens and Sparta

Democracy developed in Athens but not in all city-states. Sparta, Athens' rival, had an oligarchy ruled by a small group of warriors. They supervised a military training system for Spartan boys.

In 490 B.C., Athens and Sparta joined together to defeat the invading army of the Persian Empire under King Darius I. After that, however, the two city-states became fierce enemies.

Before You Move On
Summarize What ancient Greek ideas served as the roots of modern democracy?
establishing assemblies that represented more of the people; system of government under which all citizens voted directly for laws

Visual Vocabulary The **Acropolis** of Athens is a rocky hill that once served as the city's fortress and contained its most important temples.

ANCIENT GREEK CITY-STATES

Government in Ancient Greece
510–323 B.C.
- Limited democracy
- Oligarchy
- Monarchy
- Tyranny
- Mixed government

ONGOING ASSESSMENT
READING LAB
GeoJournal

Synthesize Use the Greek roots in the chart to form a word in English that completes each of the following sentences:

a. The form of government that represents the people is called _____ (use your own paper)

b. People empowered to enforce a city's laws are the _____ (use your own paper)

c. The ruler of a kingdom is also called a _____ (use your own paper)

SELECTED ENGLISH WORDS FORMED FROM GREEK ROOTS		
Greek Root	**Meaning**	**English Word**
demos	people	democracy
polis	city-state	policy
aristo	best	aristocracy
monos	one	monarchy
oligo	few	oligarchy

PLAN

OBJECTIVE
Identify and describe the influence of the ancient Greeks on the development of democracy.

CRITICAL THINKING SKILLS FOR SECTION 2.1

- Main Idea
- Summarize
- Synthesize
- Analyze Visuals
- Analyze Causes
- Make Inferences
- Interpret Maps

PRINT RESOURCES

Teacher's Edition Resource Bank

- Reading and Note-Taking: Track Details
- Vocabulary Practice: Definition and Details
- **GeoActivity** Analyze Primary Sources: Democracy

TECHTREK myNGconnect.com

Fast Forward!
Core Content Presentations
Teach *Roots of Democracy*

Digital Library
NG Photo Gallery, Section 2

Maps and Graphs
Online World Atlas: Ancient Greek City-States

Interactive Whiteboard
GeoActivity Analyze Primary Sources: Democracy

Also Check Out
- Graphic Organizers in **Teacher Resources**
- GeoJournal in **Student eEdition**

BACKGROUND FOR THE TEACHER

At first glance, the city-state of Athens had a more democratic government than that of the United States because its government was a direct democracy in which all citizens could vote on laws. However, citizenship in Athens was strictly limited to men who were 18 years or older and whose parents were citizens. (In 403 B.C., the age was raised to 20.) Women, foreigners, and slaves were excluded. As a result, some scholars estimate that citizens made up only 10 to 15 percent of the population.

ESSENTIAL QUESTION

How did European thought shape Western civilization?

Democracy, as discussed in Section 2.1, is one of the most widely spread forms of government in the modern world. It was first developed by the Greeks.

INTRODUCE & ENGAGE 🖥️ Digital Library

Analyze Visuals Project the images of artifacts from Athens and Sparta from the **NG Photo Gallery**. For each photograph, **ASK:**

- How was this object used?
- Does this artifact remind you of anything from your daily life?
- Does this artifact make you think that Greek culture was mostly similar to or different from U.S. culture? `0:10` **minutes**

TEACH 📊 Maps and Graphs

Guided Discussion

1. Analyze Causes Why did the Greek city-states remain independent of each other instead of merging into a single country? *(The mountains on the peninsula made communication and transportation difficult.)*

2. Make Inferences What reason might the Greeks have had for doing away with the ruling councils? *(They were controlled by aristocrats, who probably ran the government for their own benefit.)*

Interpret Maps Project the Ancient Greek City-States map from the **Online World Atlas**. Have a volunteer explain the meaning of the colors on the map legend. **ASK:**

- What bodies of water surrounded Greece? *(Ionian Sea, Sea of Crete, Aegean Sea)*
- What other city had a government system like the one in Sparta? *(Thebes)*
- Judging from the map, was Athens or Sparta more likely to be a sea power? Explain. *(Athens; It was on the coast.)* `0:15` **minutes**

DIFFERENTIATE

English Language Learners **Summarize** Allow students to work in pairs to summarize the lesson by completing the following sentences using words and phrases from the text.

- A city-state is a city and its surrounding _____.
- Each city-state was ruled by its own form of _____.
- The city-state of Athens developed a type of government called _____.
- Athens had a rival called _____.

Pre-AP **Write a Character Analysis** Have students read Odysseus's encounter with the Cyclops from the *Odyssey* and then use the Character Description chart to analyze Odysseus's character. Finally, ask students to write a brief paragraph describing what the poem reveals about the qualities the ancient Greeks admired.

CHARACTER	WHAT THE CHARACTER DOES	WHAT THIS SHOWS ABOUT THE CHARACTER

ACTIVE OPTIONS

🖼️ Interactive Whiteboard
GeoActivity

Analyze Primary Sources: Democracy Ask for volunteers to read each primary source aloud. Check to see if students have any questions about the documents. Then divide the class into small groups. Each group will work cooperatively to answer the questions about the primary sources and complete the activity. `0:15` **minutes**

EXTENSION Have groups fill out a Venn diagram comparing and contrasting democracy in ancient Greece with democracy in the United States. Suggest that they use their textbook, the two primary sources, and online sources to find the information they need. `0:20` **minutes**

On Your Feet

Numbered Heads Have students get into groups of four and number off within each group. Refer them to the chart Selected English Words Formed From Greek Roots in the Reading Lab. Tell the groups to list as many additional words as possible using the Greek roots in the chart. Allow them to use print or online dictionaries.

After ten minutes, call a number. A student from that group should read his or her group's list. Correct any errors, and then tally how many words each group found. `0:15` **minutes**

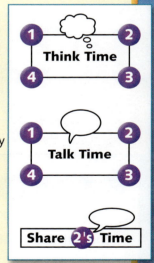

ONGOING ASSESSMENT
READING LAB ✏️ GeoJournal

ANSWERS
1. a. democratic
   b. police
   c. monarch

2.2 Classical Greece

TECHTREK
myNGconnect.com For an online map and photos of Classical Greece
Maps and Graphs | Digital Library

Main Idea Greek ideas about democracy, architecture, philosophy, and science have had a lasting influence on Western culture.

As you have learned, democracy began in ancient Greece. In 461 B.C., Pericles became the leader of Athens. His rule began a **golden age**, a period of wealth and power during which democracy developed further and Greek culture flourished.

Golden Age of Greece

Pericles had three goals for Greece. The first was to strengthen democracy. He accomplished this goal by paying citizens who held public office. This meant that even people who were not wealthy could afford to serve in government.

The leader's second goal was to expand the empire. Pericles built a strong navy and used it to increase Athens' power over the other Greek city-states.

Pericles' third goal was to make Athens more beautiful. He began rebuilding the city, including the Acropolis. Many of Athens' temples had been destroyed during the war with Persia. Pericles constructed a new temple called the Parthenon, dedicated to the goddess Athena for whom Athens was named.

Greek Achievements

The golden age of Greece was a period of extraordinary achievements. Greek architects designed temples and theaters with graceful columns. **Philosophers**, people who examine questions about the universe, searched for the truth. Socrates (SAHK ruh teez) and his student, Plato (PLAY toh), were leading philosophers.

In the sciences, the mathematician Euclid (YOO klihd) developed the principles of geometry. The physician Hippocrates (heh PAH kruh teez) changed the practice of medicine by insisting that illnesses originated in the human body and were not caused by evil spirits.

Greek Culture Spreads

Greece's golden age ended around 431 B.C., when war broke out between Athens and Sparta. The conflict, known as the Peloponnesian War, lasted 27 years and weakened both Athens and Sparta.

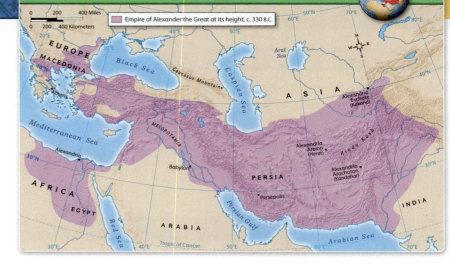

ALEXANDER'S EMPIRE, 330 B.C.

Empire of Alexander the Great at its height, c. 330 B.C.

Around 340 B.C., King Philip II of Macedonia took advantage of the weakened city-states and conquered Greece. In 334 B.C., Philip's son, **Alexander the Great**, became king and began to extend his father's empire. Alexander loved Greek culture and spread its ideas throughout the lands he conquered. Alexander died in 323 B.C. at the age of 33. The Greek ideas about democracy, science, and philosophy that he helped spread shaped the modern world.

Before You Move On
Monitor Comprehension What Greek ideas have had a lasting influence on Western culture?

ideas about democracy, architecture, philosophy, mathematics, and medicine

ONGOING ASSESSMENT
MAP LAB
GeoJournal

1. **Draw Conclusions** Study the map. Across which continents did Alexander's empire spread? What helped him unite his vast empire?

2. **Make Inferences** Based on the map, what great empires did Alexander conquer? What do these conquests suggest about Alexander?

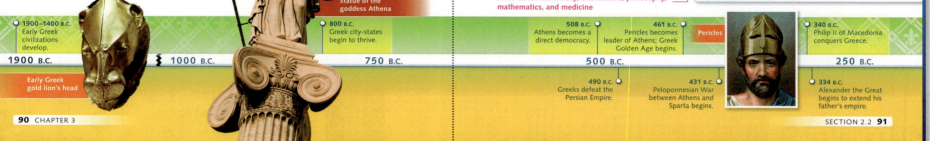

Statue of the goddess Athena

Early Greek gold lion's head

| 1900–1400 B.C. Early Greek civilizations develop. | 800 B.C. Greek city-states begin to thrive. | 508 B.C. Athens becomes a direct democracy. | 461 B.C. Pericles becomes leader of Athens; Greek Golden Age begins. | Pericles | 340 B.C. Philip II of Macedonia conquers Greece. |

1900 B.C. | **1000 B.C.** | **750 B.C.** | **500 B.C.** | **250 B.C.**

490 B.C. Greeks defeat the Persian Empire.
431 B.C. Peloponnesian War between Athens and Sparta begins.
334 B.C. Alexander the Great begins to extend his father's empire.

PLAN

OBJECTIVE Explain how the cultural achievements of classical Greece influenced the ancient and modern worlds.

CRITICAL THINKING SKILLS FOR SECTION 2.2

- Main Idea
- Monitor Comprehension
- Draw Conclusions
- Make Inferences
- Analyze Visuals
- Make Predictions
- Form and Support Opinions
- Interpret Time Lines

PRINT RESOURCES

Teacher's Edition Resource Bank

- Reading and Note-Taking: Pose and Answer Questions
- Vocabulary Practice: Word Squares
- **GeoActivity** Research Ancient Greek Contributions

TECHTREK myNGconnect.com

Fast Forward!
Core Content Presentations
Teach *Classical Greece*

Digital Library
NG Photo Gallery, Section 2

Maps and Graphs
Interactive Map Tool
Map Ancient Greece.

Also Check Out
- Online World Atlas in **Maps and Graphs**
- Graphic Organizers in **Teacher Resources**
- GeoJournal in **Student eEdition**

BACKGROUND FOR THE TEACHER

Even though Alexander the Great was Macedonian, Greek culture had a strong influence on his upbringing. His mother was a Greek princess who told him he was descended from Achilles, the great hero of the Trojan War. Alexander's father, King Philip II of Macedonia, hired the Greek philosopher Aristotle to educate his son. Aristotle, who had been a student of Plato's, taught the prince literature, history, philosophy, medicine, and science.

ESSENTIAL QUESTION

How did European thought shape Western civilization?

Section 2.2 discusses Greek achievements and culture, pointing out ideas about drama, philosophy, and geometry that still influence Western culture.

INTRODUCE & ENGAGE 🖳 Digital Library

Analyze Visuals Project the images of Greek gods and goddesses from the NG Photo Gallery. After showing the photographs, **ASK:** Do you know any stories about the Greek gods and goddesses? *(Responses will vary.)* Does modern culture still contain references to these gods and goddesses? *(Responses will vary, but students might mention cultural references such as the* Percy Jackson and the Olympians *series of novels and companies named after gods and goddesses.)*

`0:10` **minutes**

TEACH

Guided Discussion

1. **Make Predictions** Discuss with students the likely impact of Pericles' decision to pay public officials. **ASK:** What would be the long-term impact of Pericles' policy? *(Government would become more representative of a broader range of economic classes.)*

2. **Form and Support Opinions** Review the achievements introduced during the golden age of Greece. **ASK:** Which of the Greek achievements do you think has the biggest effect on your life today? *(Possible response: I think Hippocrates' changes to medicine have the biggest impact because doctors still believe that illness originates in the human body.)*

Interpret Time Lines Help students interpret the time line by reminding them that with B.C. dates, the larger the number, the further back it is in time. Also remind them that the jagged line to the left of 1000 B.C. indicates a break in the time line because of a long passage of years. **ASK:** About how many years after the beginning of city-states did Athens become a democracy? *(about 300)* What event on the time line caused Greek city-states to lose their independence, and when did it occur? *(Philip II of Macedonia conquered Greece in 340 B.C.)* `0:15` **minutes**

DIFFERENTIATE

Inclusion Work in Pairs Pair students who have visual or learning disabilities with partners who are proficient readers. Have the proficient student read the section titled "Greek Achievements" aloud. After each sentence, the reader should stop so the partner can discuss whether an achievement was mentioned. If it was, the pair should list it on an Idea Web similar to the one shown here. Explain that students may add more circles to the Idea Web if needed.

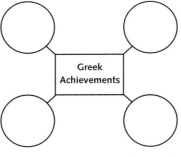

Greek Achievements

Pre-AP Hold a Panel Discussion Have groups of students research Pericles and the changes he made. Then have them present a panel discussion, explaining the most important of Pericles' reforms and discussing whether he was a good leader. Encourage students to cover the following points in their discussion:

- changes to standards of citizenship
- representation within the assembly
- changes to the judiciary

ACTIVE OPTIONS

🗺 Interactive Map Tool

Map Ancient Greece

PURPOSE Locate places in ancient Greece

SET-UP

1. Open the **Interactive Map Tool**, set the "Region" to Europe, set the "Map Mode" to Outline, and click on Greece to open up a more detailed map.

2. Under "Other Features," deselect Points of Interest, Large Cities, and Small Cities.

3. Click on the DRAWING TOOLS tab.

ACTIVITY

1. Ask a volunteer to use the Label Tool to label the island of Crete. Ask a second volunteer to label the island of Rhodes.

2. Click on the MARKERS tab. Ask another volunteer to place a marker on the location of the city of Sparta and then use the Label Tool to label it.

3. Have additional volunteers locate and label each of the following cities: Corinth, Delphi, Messenia, Olympia, Thebes. `0:20` **minutes**

On Your Feet

Turn and Talk on Topic Have students form four lines. Give each line the same topic sentence: *The ancient Greeks accomplished many things during their golden age.* Tell the groups to build a paragraph on that topic by having each student in the line add one sentence about a different achievement of ancient Greece. Finally, have groups present their paragraphs to the class, with each student reading his or her sentence. `0:15` **minutes**

ONGOING ASSESSMENT

MAP LAB 📋 GeoJournal

ANSWERS

1. Europe, Asia, and Africa; He united the empire by spreading Greek culture throughout it.

2. He conquered the Persian Empire and Mesopotamia. These conquests suggest that Alexander was bold, brave, and a good general.

2.3 The Republic of Rome

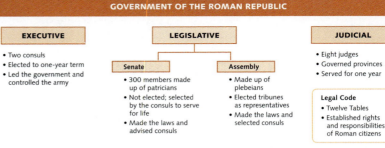

TECHTREK
myNGconnect.com For an online map of ancient Rome and photos of Roman ruins
Maps and Graphs
Digital Library

Main Idea The Roman Republic created a form of government that Europe and the West would later follow.

Around 1000 B.C., the peninsula of Italy was dotted with hundreds of small villages. According to an ancient legend, two brothers named Romulus and Remus founded Rome in 753 B.C. The brothers were said to be the children of a god and to have been raised by a wolf.

The Beginnings of Rome

Archaeologists actually believe that people known as the Latins founded Rome around 800 B.C. They came from a region of Italy called Latium and lived on Rome's seven steep hills, which provided protection from enemy attack. The Tiber River, which flows through Rome, provided water for farming and a route for trade. Over time, Rome developed into a wealthy city-state.

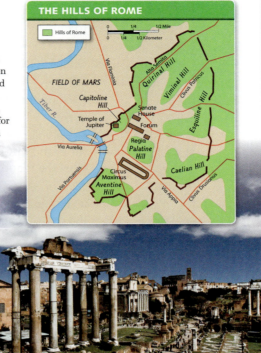

Critical Viewing The Roman Forum contained the ancient city's most important buildings, including the Senate. In what ways does this photo reflect Rome's former glory?

The ruins show that the buildings were large, imposing, and of beautiful architecture.

A Republic Forms

Around 600 B.C., the Etruscans, a people from northern Italy, conquered Rome. One Etruscan king named Tarquin was a brutal tyrant. In 509 B.C., the Romans rebelled against him, and Roman leaders began to create a republic. A **republic** is a form of government in which the people elect officials who govern according to law.

THE HILLS OF ROME

Hills of Rome

FIELD OF MARS

Tiber R.

Capitoline Hill
Temple of Jupiter
Senate House
Forum
Regia
Palatine Hill
Circus Maximus
Aventine Hill

Alta Semita
Quirinal Hill
Viminal Hill
Clivus Patricus
Esquiline Hill
Caelian Hill
Clivus Drusianus

Via Flaminia
Via Aurelia
Via Portuensis
Via Appia

GOVERNMENT OF THE ROMAN REPUBLIC

EXECUTIVE	LEGISLATIVE		JUDICIAL
• Two consuls • Elected to one-year term • Led the government and controlled the army	**Senate** • 300 members made up of patricians • Not elected; selected by the consuls to serve for life • Made the laws and advised consuls	**Assembly** • Made up of plebeians • Elected tribunes as representatives • Made the laws and selected consuls	• Eight judges • Governed provinces • Served for one year **Legal Code** • Twelve Tables • Established rights and responsibilities of Roman citizens

Two main classes of people lived in Rome at this time. The **patricians** were mostly wealthy landowners. The **plebeians** were mostly farmers. At first, only the patricians could take part in government. They controlled the Senate and made laws.

In 490 B.C., plebeians gained the right to form an assembly and elect legislative representatives called tribunes. The assembly made laws and elected the consuls, the two executive officials who led the government for a year at a time. One consul could **veto**, or reject, a decision made by the other consul.

The judicial branch was made up of eight judges who served for one year. These judges oversaw the lower courts and governed the provinces.

Around 450 B.C., the government published the Twelve Tables. These were bronze tablets that set down the rights and responsibilities of Roman citizens. At this time, only adult male landowners born in Rome were citizens. Roman women were citizens but could not vote or hold office.

The Roman Way

Citizens of Rome believed in values that were known as the Roman Way. These values included showing self-control,

a legal code, a government with three branches, a system with some checks and balances, civic values of doing one's duty and pledging loyalty to Rome

working hard, doing one's duty, and pledging loyalty to Rome. The Roman Way helped to unite all Roman citizens.

The Romans applied these values as the Republic began to conquer new lands and expand. During the second century B.C., Rome defeated the empire of Carthage in northern Africa. By 100 B.C., Rome controlled most of the lands around the Mediterranean. About this time, tensions began to grow between patricians and plebeians. These tensions triggered a war between the two groups. The war set the stage for the end of the Roman Republic—and the birth of the Roman Empire.

Before You Move On

Summarize What structures, laws, and values made up the government of the Roman Republic?

ONGOING ASSESSMENT
DATA LAB GeoJournal

1. **Interpret Charts** According to the chart, members of the Senate were selected by the consuls. However, the assembly elected the consuls. In what way did this arrangement help to control the Senate's power?

2. **Turn and Talk** Study the chart and consider what you know about the U.S. government. Then, with a partner, compare and contrast the two systems of government.

SECTION 2.3 **93**

PLAN

OBJECTIVE Compare the structure of the U.S. government with that of the Roman Republic.

CRITICAL THINKING SKILLS FOR SECTION 2.3

- Main Idea
- Summarize
- Interpret Charts
- Analyze Causes
- Synthesize
- Compare
- Interpret Maps

PRINT RESOURCES

Teacher's Edition Resource Bank

- Reading and Note-Taking: Find Main Idea and Details
- Vocabulary Practice: Compare/Contrast Paragraph
- **GeoActivity** Compare Greek and Roman Governments

TECHTREK myNGconnect.com

Fast Forward!
Core Content Presentations
Teach *The Republic of Rome*

Maps and Graphs
Online World Atlas: The Hills of Rome

Connect to NG
Research Links

Interactive Whiteboard
GeoActivity Compare Greek and Roman Governments

Digital Library
NG Photo Gallery, Section 2

Also Check Out
GeoJournal in Student eEdition

BACKGROUND FOR THE TEACHER

No written accounts of early Roman history have survived. According to legend, seven kings ruled from 753 to 509 B.C. Six of these kings were just, kindly rulers while the seventh, Lucius Tarquinius Superbus, was so despotic that a popular uprising overthrew him. However, modern scholars believe that the accounts are probably stories used to explain the transition from monarchy rather than accurate histories.

ESSENTIAL QUESTION

How did European thought shape Western civilization?

The Romans created a republic with three branches of government and based on written laws. Section 2.3 shows how Roman law is the basis for the legal code of many Western countries and how Roman values have influenced those of Western countries.

INTRODUCE & ENGAGE

Team Up: Brainstorm Divide the class into small groups. Assign half the groups to represent wealthy landowners and the other half to represent farmers. Explain that the government of their land is controlled by the landowners. Have groups use a chart like the one below to list their group's views about whether the government should change. In the support column, they should list their reasons. Then have groups representing both sides of the argument discuss their viewpoints. Ask students to list the opposing group's views in the third column during the discussion. After the discussion is over, tell students that they will read how ancient Rome dealt with this situation. `0:15` minutes

VIEWPOINT	SUPPORT	OPPOSING VIEWPOINT

TEACH Maps and Graphs

Guided Discussion

1. **Analyze Causes** What geographic features influenced the location of the city of Rome? *(Steep hills provided a safe site for villages, and the Tiber River provided water.)*

2. **Synthesize** How did the structure of the legislative branch of government reflect the class system in Rome? *(There were two main classes: patricians and plebeians. Each one had a house in the legislative branch. The patricians had the Senate, and the plebeians had an assembly.)*

3. **Compare** Discuss some of the responsibilities of being a good citizen. **ASK:** How do the values of the Roman Way compare to the qualities of good citizenship today? Explain. *(Possible response: They are similar. Good citizens work hard to support themselves, show self-control in obeying the law, do their duty such as voting and serving in the military, and pledge allegiance to the U.S. flag.)*

Interpret Maps Point out The Hills of Rome map in the lesson or project it from the **Online World Atlas**. Have students choose partners and work together to write three questions about the map. They should record the answers on a separate sheet of paper. Then have each pair exchange their questions with another pair and try to answer the new set. `0:20` minutes

DIFFERENTIATE Connect to NG

Striving Readers Summarize Read the lesson aloud while students follow along in their books. At the end of each paragraph, ask students to summarize what you read in a sentence. Allow them time to write the summary on their own paper.

Pre-AP Write a Military Biography Have students use the **Research Links** to research one of the military leaders from the Punic Wars between Rome and Carthage, such as Hannibal, Scipio Africanus the elder, or Scipio Africanus the younger. Then ask them to write a short biography and share it with the class.

Hannibal

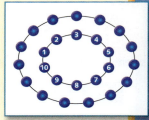

2.4 The Roman Empire

TECHTREK
myNGconnect.com For an online map of the
Roman Empire and photos of Roman architecture

Maps and Graphs Digital Library

Main Idea The Roman Empire was one of the largest in history and left a legacy in technology and language.

As you have read, tensions grew between the patricians and plebeians not long after Rome defeated Carthage. The Roman soldiers had brought back great wealth from the conquered territory. They used their wealth to buy large plots of farmland. Small farmers could not compete with them, and the gap between rich and poor widened. In 88 B.C., war erupted between the patricians and plebeians.

Creation of the Empire

After years of war, a general named **Julius Caesar** rose to power and became the sole ruler of Rome in 46 B.C. Caesar started projects to help the poor and tried to re-establish order in Rome, but he developed powerful enemies. In 44 B.C., a small group of senators stabbed him to death.

Octavian, Caesar's nephew, fought in the long, internal war for power that followed Caesar's death. Octavian won, but the war put an end to the Roman Republic. Calling himself **Augustus**, which means "honored one," Octavian became the ruler of the Roman Empire in 27 B.C. His rule began a period called the *Pax Romana*, which is Latin for "Roman peace."

Rome's Decline

For about 500 years, the Roman Empire was the most powerful in the world and extended over three continents. Around A.D. 235, however, Rome had a series of poor rulers. In addition, German tribes, whom the Romans called **barbarians**, began invading the empire from the north.

In 330, Emperor Constantine moved the capital of the weakened empire from Rome to Byzantium, in present-day Turkey, and renamed the city Constantinople. He also made Christianity lawful throughout the empire. **Christianity**, which is based on Jesus' life and teachings as described in the Bible, began in the Roman Empire.

In 395, the empire was divided into an Eastern Empire and a Western Empire with two different emperors. In 476, invaders overthrew the last Roman emperor and ended the Western Empire.

Rome's Legacy

The Western Empire fell, but Rome left the world a great legacy, or heritage. For example, Roman engineers built a network of roads that connected the empire. Many of the roads are still in use.

The engineers also developed the arch and used it to construct buildings and **aqueducts**, which carried water to parts of the empire. Latin, the language of Rome, became the basis for Romance languages, such as Spanish and Italian. Many English words have Latin roots.

Before You Move On
Summarize Describe the Roman Empire's rise, fall, and legacy. It rose with Augustus as its leader. It flourished for hundreds of years until invaders finally overthrew in A.D. 476. It left behind a legacy in technology, architecture, and language.

THE ROMAN EMPIRE AT ITS HEIGHT, A.D. 200

Expansion of the Roman Empire
Roman Republic in 264 B.C.
Roman Empire at its height, c. A.D. 200

0   200   400 Miles
0   200   400 Kilometers

ONGOING ASSESSMENT
MAP LAB GeoJournal

1. **Interpret Maps** According to the map, over which continents did the Roman Empire extend? What challenges might the size of the empire have presented to its rulers?

2. **Location** Find Byzantium on the map. Why do you think Constantine chose this location to become the capital of the Eastern Empire?

88 B.C.
Civil wars begin.

44 B.C.
Caesar is murdered.

27 B.C.
Augustus becomes emperor.

100 B.C.   A.D. 100   A.D. 300   A.D. 500

Roman coin showing Julius Caesar

46 B.C.
Julius Caesar becomes emperor.

Statue of Augustus

A.D. 330
Constantine moves the capital from Rome to Byzantium.

A.D. 395
Empire is divided into an Eastern Empire and a Western Empire.

A.D. 476
Rome falls to invaders.

Painting of German invader

94 CHAPTER 3

PLAN

OBJECTIVE Analyze the rise and fall of the Roman Empire and the impact of Roman culture on Western civilization.

CRITICAL THINKING SKILLS FOR SECTION 2.4

- Main Idea
- Summarize
- Interpret Maps
- Make Inferences
- Identify Problems
- Find Details
- Create Time Lines

PRINT RESOURCES

Teacher's Edition Resource Bank

- Reading and Note-Taking: Sequence Events
- Vocabulary Practice: W-D-S Triangles
- **GeoActivity** Analyze the Roots of Modern Languages

TECHTREK myNGconnect.com

Fast Forward!
Core Content Presentations
Teach *The Roman Empire*

Connect to NG
Research Links

Interactive Whiteboard
GeoActivity Analyze the Roots of Modern Languages

Also Check Out
- NG Photo Gallery in **Digital Library**
- Online World Atlas in **Maps and Graphs**
- GeoJournal in **Student eEdition**

BACKGROUND FOR THE TEACHER

During his reign, Augustus Caesar reorganized the government, reforming the administrative structures and beginning the first Roman civil service. These improvements fostered communication and trade and thus helped hold the empire together. He also embarked on a massive public works project of improving Rome. According to the historian Suetonius, Augustus boasted that he found the city built of brick and left it built of marble.

ESSENTIAL QUESTION

How did European thought shape Western civilization?

The Roman Empire spread Roman technology, engineering advances, and artistic styles through much of Europe, as discussed in Section 2.4. Rome also spread the Latin language, which became the basis of Romance languages.

INTRODUCE & ENGAGE

Activate Prior Knowledge Ask a volunteer to name the form of government Rome adopted after overthrowing Etruscan rule. *(a republic)* Remind students that during the republic, Rome expanded to control most of the Italian Peninsula. Ask students to predict how conquests of foreign lands even farther away from Rome might change the structure of the government. `0:10` minutes

TEACH 🖱 Connect to NG

Guided Discussion

1. **Make Inferences** Help students understand why Julius Caesar was murdered. **ASK:** Why do you think the senators who killed Julius Caesar wanted him dead? *(Possible response: He wanted to help the poor, a policy that threatened the status of the rich patricians. He became sole ruler of Rome, and Roman tradition was against kings.)*

2. **Identify Problems** What weakened the Roman Empire? *(Rome had a series of poor rulers, and German tribes began invading the northern part of the empire.)*

3. **Find Details** What are some modern languages influenced by Latin? *(English, French, Italian, Spanish)*

Create Time Lines Have small groups work together to add topics to the time line in their textbook. First have them choose a topic from Roman history, such as emperors, wars, or culture. Have groups research their topic at the library or by using the **Research Links**, choose an event, and write a text box for it that includes citations for their sources. `0:20` minutes

DIFFERENTIATE 📱 Connect to NG

English Language Learners Create a Word Map Give students these related words: *empire, emperor, imperial.* Have them work in pairs to create a Word Map for each one like the one below. Students should write a sentence for each of the three words and then exchange with his or her partner to make sure that the words are used correctly. As a challenge, have students look up each word and find the Latin root. **ASK:** Is each word related to the same root? *(Yes, all the words are related to imperare.)*

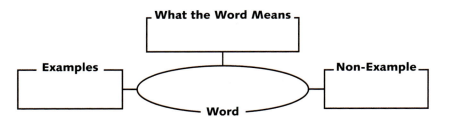

Gifted & Talented Do a Roman Home Makeover Have students use the **Research Links** to learn how Roman homes were decorated with murals and mosaics. Ask them to find out what materials were used to make the mosaics, who typically owned them, and what purpose the mosaics might have served. Then have students paint or draw a mural on a large sheet of paper to decorate part of the classroom in the style of the Romans.

Roman Mosaic

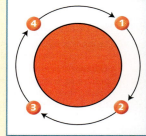 

2.5 Middle Ages and Christianity

TECHTREK
myNGconnect.com For portrayals of the Crusades in fine art

Digital Library

Main Idea The Roman Catholic Church and the feudal system influenced Western Europe during the Middle Ages.

After the fall of the Roman Empire, Western Europe entered a period known as the **Middle Ages**, which lasted from about 500 to 1500. During this period, Western Europe consisted of numerous kingdoms. Castles, like those on the Rhine River in Germany, served as defensive fortresses. The Roman Catholic Church helped unite people during the Middle Ages, and the feudal system provided a social structure.

The Roman Catholic Church

In 1054, Christianity officially divided into two parts: the Roman Catholic Church in Western Europe and the Eastern Orthodox Church in Eastern Europe. The Roman Catholic Church was the center of life for most people in Western Europe. It cared for the sick, provided education, and helped preserve books and learning.

The Church also played a leading role in government. It collected taxes, made its own laws, and waged wars. In 1096, the Church began a series of **Crusades**. These were military expeditions undertaken to take back holy lands in Southwest Asia from Muslim control. The Crusades cost many lives and ended in 1291.

The Feudal System

The many kingdoms of Western Europe were often at war. From about 400 to 800, a German group called the Franks stopped the fighting and unified most of Western Europe. Their most important leader was Charlemagne (SHAHR luh mayn).

When Charlemagne died in 814, warfare between the kingdoms returned and Western Europe again became divided. To provide security for each kingdom, the feudal system developed. The **feudal system** was a social structure that was organized like a pyramid. At the top was a king who owned vast territory. Beneath the king were lords, powerful noblemen who owned land. The lords gave pieces of their land to vassals, who pledged their loyalty and service to the lord. Some vassals also served as knights, who were warriors on horseback.

Each lord lived on an estate called a manor, which functioned as a small village. **Serfs**, who farmed the lord's land in return for shelter and protection, were at the bottom of the pyramid. Serf families dwelt in small huts on the manor and gave most of the crops they grew to their lord.

The Growth of Towns

In time, the growth of towns helped end the feudal system. Trade and businesses developed, and people began to leave the manors. A deadly disease called the bubonic plague, which swept through Europe in 1347, also weakened the feudal system. The plague killed millions and greatly reduced the workforce in the towns. Desperate for workers, employers offered higher wages. As a result, farmers left the country to seek the higher-paying jobs in the towns.

Before You Move On

Summarize In what ways did the Roman Catholic Church and the feudal system influence Western Europe during the Middle Ages? The Church united people and dominated their lives. The feudal system provided a social structure.

MANOR IN THE MIDDLE AGES

This illustration shows a simplified view of a feudal manor. Rolling fields and farmland lay outside its walls.

The church was the center of life on the manor.

The lord lived in relative safety and ease in his castle.

Serfs lived in tiny huts with dirt floors.

Guards protected the manor from rival lords.

ONGOING ASSESSMENT

VIEWING LAB GeoJournal

1. **Interpret Visuals** What details in the illustration suggest the measures taken to protect those who lived in the manor?
2. **Make Inferences** Notice the position of the church in the manor. Why might it have been positioned near the lord's castle?
3. **Compare and Contrast** Based on the illustration and what you have read, in what ways did life probably differ for lords and serfs?

SECTION 2.5 97

PLAN

OBJECTIVE Draw conclusions about life in the Middle Ages by analyzing the Roman Catholic Church and the feudal system.

CRITICAL THINKING SKILLS FOR SECTION 2.5

- Main Idea
- Summarize
- Interpret Visuals
- Make Inferences
- Compare and Contrast
- Identify Main Ideas
- Analyze Causes

PRINT RESOURCES

Teacher's Edition Resource Bank

- Reading and Note-Taking: Summarize Information
- Vocabulary Practice: Vocabulary Pyramid
- **GeoActivity** Categorize Effects of the Crusades

TECHTREK myNGconnect.com

Fast Forward!
Core Content Presentations
Teach *Middle Ages and Christianity*

Connect to NG
Research Links

Digital Library
NG Photo Gallery, Section 2

Also Check Out
- Charts & Infographics and Graphic Organizers in **Teacher Resources**
- GeoJournal in **Student eEdition**

BACKGROUND FOR THE TEACHER

The first castles were little more than fortresses atop a high mound with an encircling ditch. Being elevated gave the defenders an advantage over advancing troops. By the 1000s, baileys, or grounds between encircling walls, were added to castle defenses. The outermost wall was often defended by a drawbridge over the moat and a portcullis, or grate, that could close off the gateway. By the late 1100s, the keep, or fortress, of the castle was often built against a cliff to block off one line of approach.

ESSENTIAL QUESTION

How did European thought shape Western civilization?

Section 2.5 discusses how Christianity became a major influence in European life and politics and later in European colonies in other lands.

INTRODUCE & ENGAGE

Connect to Modern Life Ask students to name various ways that people try to protect their homes today, and make a class list of their responses. *(Possible responses: guard dogs, fences, alarm systems, security guards)* Remind students that the Western Roman Empire fell in 476 and with it went the strong central government that had kept peace and order. Ask students to predict what people did to protect themselves once the Roman army was gone. `0:10` minutes

TEACH

Guided Discussion

1. **Identify Main Ideas** Have students read the first two headings in the lesson. **ASK:** What unified Western Europe after the Western Roman Empire ended? *(the Roman Catholic Church)* What system did Europeans develop to provide security in the Middle Ages, and how did it work? *(Europeans developed the feudal system. It was a social structure in which lords gave land to vassals who pledged loyalty and service, including fighting as warriors.)*

2. **Analyze Causes** Now have students read the last heading in the lesson. **ASK:** What led to the growth of towns after the year 1000? Draw a Cause-and-Effect chart on the board like the one shown below to list their ideas. *(Farmers began to use new technology that helped them grow more crops. They took their extra crops to markets, and towns grew up in those places.)*

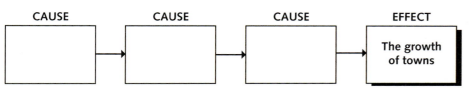

CAUSE → CAUSE → CAUSE → EFFECT: The growth of towns

MORE INFORMATION

The Crusades In 1071, Muslim Turks defeated Byzantine Christians at Manzikert, Turkey. As the Turks continued to gain territory, the Byzantine emperor asked Pope Urban II for help. In 1095, Urban called for Western Christians to go on a crusade. The First Crusade took Muslims by surprise, and the Crusaders succeeded in capturing Jerusalem and establishing a Latin kingdom. However, the famous sultan Saladin retook Jerusalem in 1187, and subsequent Crusades failed to halt Muslim expansion.

DIFFERENTIATE 🔵 Connect to NG

Inclusion **Describe Details in Drawings** Pair students who are visually impaired with students who are not. Ask the latter to be their partners' "eyes" and describe the details in the Manor in the Middle Ages drawing. Then have the partners work together to answer the questions in the Viewing Lab.

Pre-AP **Create Maps** Have students use the **Research Links** to learn the extent of Charlemagne's empire and the way it was divided by the Treaty of Verdun. Have them create a map showing both the boundaries of the empire and its later division. Allow them to display their maps in the classroom.

ACTIVE OPTIONS

📷 NG Photo Gallery

Analyze Visuals Have students view photographs of art and artifacts of the Middle Ages from the **NG Photo Gallery** and answer the following questions with a partner.

ILLUSTRATED MANUSCRIPT
- What skills did a person need to create an illustrated manuscript?
- Why do you think medieval books were illustrated?

BUST OF CHARLEMAGNE
- What materials do you think this is made of?
- What impression did the sculptor want to give of Charlemagne?

THE CRUSADES
- What does this photo tell you about warfare in the Middle Ages?

THE PLAGUE
- Who are the figures in this photo?
- What emotion does this photo convey?

KNIGHT
- What do you think are the disadvantages of wearing armor like this? `0:15` minutes

On Your Feet

Three-Step Interview
Have students choose a partner. One student should interview the other on the question, *Would you have liked to live in the Middle Ages? Why or why not?*
Then they should reverse roles. Finally, each student should share the results of his or her interview with the class. `0:20` minutes

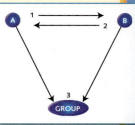

ONGOING ASSESSMENT
VIEWING LAB 📓 GeoJournal

ANSWERS
1. the walls all around the village, guard towers with guards, the gate in the manor behind which people could be safer
2. The church may have been near the castle because both of them represented power and safety—government—to the people in the village.
3. Responses will vary, but lords probably ate rich foods like meat and butter, had nicer clothes, and lived in a warmer house. The serfs probably lived in drafty huts, wore rough and ragged clothes, and had only basic foods to eat.

2.6 Renaissance and Reformation

TECHTREK
myNGconnect.com For examples of Renaissance art

Digital Library

Main Idea Both the Renaissance and the Reformation brought great change to Europe.

You have learned that the growth of towns in Western Europe helped put an end to the feudal system. The Roman Catholic Church also began to lose some of its power at this time. As these key structures of the Middle Ages weakened, the **Renaissance** began to take hold. The Renaissance was a rebirth of art and learning that started in Italy in the 1300s and had spread through Europe by 1500.

The Renaissance

Several other factors led to the Renaissance. Increased trade in the growing towns brought some Italian merchants great wealth. This wealth allowed them to buy the artists' work.

As you have learned, the Western part of the Roman Empire fell in 476. The Eastern part continued and became known as the Byzantine Empire. The fall of this empire in 1453 also advanced the Renaissance. Scholars from the empire came to Italy, bringing with them ancient writings of the Greeks and Romans. Studies of these works encouraged humanism, which focuses on human rather than religious values. In addition,

a German inventor named **Johannes Gutenberg** developed a printing press in 1450 that printed many books in a short amount of time. Soon, more people in Europe had access to knowledge—including the new humanist ideas.

The result was an explosion in art, architecture, and literature. Renaissance artists such as Leonardo da Vinci, Michelangelo (my kuhl AN juh loh), and Raphael (rah fee ELL) used perspective to make a painting look as if it had three dimensions. Architects used elements of ancient Greek and Roman design to create churches and buildings. Writers wrote in the vernacular, the language spoken in a particular region. For example, Dante Alighieri (ah lah GYER ee), who bridged the Middle Ages and Renaissance, wrote his work, *The Divine Comedy*, in Italian, not Latin.

The Reformation

Meanwhile, some people began looking more critically at the Church. **Martin Luther**, a monk in Germany, was shocked by the corrupt practices of some priests. To raise funds, they often sold indulgences, which relaxed the penalty for sin.

Critical Viewing *The School of Athens* by Italian artist Raphael portrays important ancient Greek philosophers and scientists. Why do you think the artist chose to celebrate ancient Greece? **because ancient Greece witnessed great advances in learning and culture**

The Renaissance helped shift the focus from religious values to human values. The Reformation ended some corrupt practices in the Catholic Church and led to the formation of Protestantism.

In 1517, Luther wrote the 95 Theses, in which he objected to such practices, and nailed them to a church door. Luther's actions started the **Reformation**, the movement to reform Christianity. Over time, people founded Protestant churches. The term comes from the word *protest*.

In response, the Church began a reform movement called the **Counter-Reformation** and placed more emphasis on faith and religious behavior. Nonetheless, the conflict between Catholics and Protestants would continue for the next 300 years.

Before You Move On

Monitor Comprehension What changes did the Renaissance and Reformation bring to European culture and society?

ONGOING ASSESSMENT
READING LAB
GeoJournal

1. **Summarize** What was the Renaissance?

2. **Analyze Cause and Effect** In what ways did the development of Gutenberg's printing press help spread humanist ideas?

3. **Draw Conclusions** What humanist ideas might have led people to look at the Roman Catholic Church more critically?

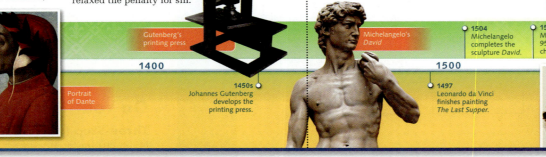

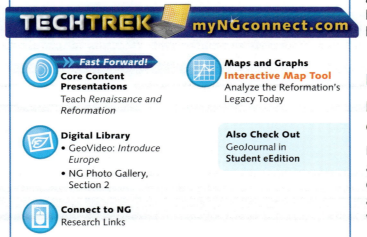

| 1300s | 1308 | 1400 | 1450s | 1497 | 1500 | 1504 | 1517 | 1600 |

- **1300s** Renaissance begins in Italy.
- **1308** Dante starts writing *The Divine Comedy* in Italian.
- Portrait of Dante
- Gutenberg's printing press
- **1450s** Johannes Gutenberg develops the printing press.
- Michelangelo's *David*
- **1497** Leonardo da Vinci finishes painting *The Last Supper*.
- **1504** Michelangelo completes the sculpture *David*.
- **1517** Martin Luther nails his 95 Theses to the door of a church in Wittenberg, Germany.
- In this illustration, Martin Luther posts the 95 Theses.

SECTION 2.6 **99**

PLAN

OBJECTIVE Analyze the cultural changes that took place in Europe during the Renaissance and Reformation.

CRITICAL THINKING SKILLS FOR SECTION 2.6

- Main Idea
- Monitor Comprehension
- Summarize
- Analyze Cause and Effect
- Draw Conclusions
- Analyze Visuals

PRINT RESOURCES

Teacher's Edition Resource Bank

- Reading and Note-Taking: Analyze Cause and Effect
- Vocabulary Practice: Definition Chart
- **GeoActivity** Map the Protestant Reformation

TECHTREK myNGconnect.com

Fast Forward!
Core Content Presentations
Teach *Renaissance and Reformation*

Digital Library
- GeoVideo: *Introduce Europe*
- NG Photo Gallery, Section 2

Connect to NG
Research Links

Maps and Graphs
Interactive Map Tool
Analyze the Reformation's Legacy Today

Also Check Out
GeoJournal in Student eEdition

BACKGROUND FOR THE TEACHER

The Medici were some of the most famous patrons of the arts during the Renaissance. They were a family of bankers who gained so much power that they came to rule Florence. Lorenzo de' Medici, who lived from 1449 to 1492, used his wealth and status to support scholarship and the arts. For example, he allowed the young Michelangelo to live and work in his home. For his achievements as ruler and patron, he became known as Lorenzo the Magnificent.

ESSENTIAL QUESTION

How did European thought shape Western civilization?

Renaissance writers introduced new forms and styles and new ways of thinking. Section 2.6 describes how during the Reformation, new kinds of Christianity arose, and the relationship between church and state was altered.

INTRODUCE & ENGAGE 📺 Digital Library

GeoVideo: *Introduce Europe* Show students the portion of the video that deals with the Renaissance. Invite students to share their impressions of the art and period.

Four Corner Activity: Changing Society Post the four signs as described below. Tell students to suppose that they are part of a cutting-edge movement that wants to change society for the better. Tell them to choose which of the four influences they will rely on as the inspiration for those changes, move to the appropriate corner, and explain their reasons. `0:15` minutes

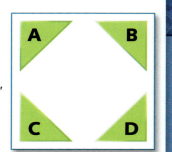

- **A Foreign Culture**—the art, customs, language, and food from other lands
- **B Religious Tradition**—the core teachings and laws of a particular religion
- **C Ancient Learning**—the wisdom of an earlier civilization
- **D New Technology**—an invention that improves people's lives in some way

TEACH 📺 Digital Library

Guided Discussion

1. Draw Conclusions What events helped spread learning during the Renaissance? *(The fall of the Byzantine Empire caused scholars to carry ancient learning west; the invention of the printing press made it easier to print books.)*

2. Analyze Effects What unintended effects did Martin Luther's actions have? *(His actions led to a split in the Church and 300 years of religious conflict.)*

Analyze Visuals Project the photos of Renaissance art from the **NG Photo Gallery**. After showing the photographs, **ASK:**

1. Have you ever seen any of the works of art before? If so, where?
2. What is the subject matter of these works?
3. What words would you use to describe the style of these works of art? `0:15` minutes

DIFFERENTIATE 🖱 Connect to NG

English Language Learners **Use Sentence Strips** Choose a paragraph from the lesson and make sentence strips out of it. Read the paragraph aloud, having students follow along in their books. Have students close their books and give them the set of sentence strips. Students should put the strips in order and then read the paragraph aloud.

Gifted & Talented **Demonstrate Perspective** Have students use the **Research Links** to learn how artists use linear perspective in paintings. Then ask them to make an informational poster by taking a photocopy of a Renaissance work of art, drawing the perspective lines over the image, and then mounting the diagram on a piece of poster board with an explanation beneath. Ask them to use the resulting poster to teach perspective to the rest of the class.

DaVinci's *Mona Lisa*

ACTIVE OPTIONS

Interactive Map Tool

Analyze the Reformation's Legacy Today
PURPOSE Analyze how the Reformation caused permanent religious divisions in Europe

SET-UP
1. Open the **Interactive Map Tool**, set the "Region" to Europe, and set the "Map Mode" to Topographic.
2. Under "Human Systems—Populations & Culture," turn on the Major Religions layer. Slide the transparency to 35 percent.

ACTIVITY
Ask students to describe the religious divisions in Europe today. *(Eastern Europe is Eastern Orthodox. Northern Europe is Protestant. Southern Europe and part of Ireland is Catholic. A small region of central Europe is mixed Christian.)* **ASK:** How does the distribution of religion in Europe today reflect the impact of the Reformation? *(Possible response: The Reformation began in Germany, and lands near Germany are still Protestant, while lands closer to Italy, where the pope lived, remained Catholic.)* `0:15` minutes

On Your Feet

Half and Half Divide the class in half. Ask one half to discuss the influence of the Renaissance on European culture and the other half to discuss the influence of the Reformation. Then have each team present a report in a way that is reflective of their topic. For example, the Renaissance team might use art in their presentation. `0:15` minutes

Performance Assessment

Quiz Each Other Divide the class into an even number of groups. Have each group create a matching quiz about important people, events, and terms from early European history. Then have pairs of groups exchange papers, take the quizzes, and then check each other's answers. Go to **myNGconnect.com** for the rubric.

ONGOING ASSESSMENT
READING LAB 📝 GeoJournal

ANSWERS
1. a period of rebirth in arts and learning
2. Books could be printed cheaply. They made the ideas of the Greeks, Romans, and European thinkers available to more people.
3. People were questioning corruption in the Church given the new emphasis on human values and achievements.

3.1 Exploration and Colonization

TECHTREK
myNGconnect.com For an online map of European colonization in Africa, Asia, and the Americas
Maps and Graphs

Main Idea To expand trade, Europeans explored Africa, Asia, and the Americas and established colonies on all three continents.

Around 1415, Prince Henry of Portugal decided that he would send explorers to Africa to establish new trade routes. Henry, who became known as Prince Henry the Navigator, founded a **navigation** school. The school taught sailors about mapmaking and shipbuilding and marked the beginning of the Age of Exploration.

European Exploration

Portugal was the first of many European countries to sponsor voyages of exploration. Europeans wanted to find gold and establish trade with Asia to obtain spices, silk, and gems. They also wanted people in other lands to **convert**, or change their religion, to Christianity.

The voyages were filled with danger. Explorers often sailed for months in ships that were small and not always able to withstand strong storms at sea. The men also faced disease and attacks by native peoples. Furthermore, the explorers were traveling to unknown lands. Mapmakers often marked unexplored places with the phrase "Here be dragons."

Nonetheless, Portuguese explorers such as Bartolomeu Dias and Vasco da Gama sailed along the coast of Africa in the late 1400s to open up trade with Asia. Italian explorer Christopher Columbus uncovered a "new world"—the continents of North America and South America—in 1492. In the 1530s, Jacques Cartier (kahr TYAY) explored parts of North America for France. An Englishman, Sir Francis Drake, sailed around the world in 1577.

EARLY COLONIZATION OF AFRICA, ASIA, AND THE AMERICAS

European Colonies c. 1750
- Britain and possessions
- Spain and possessions
- Portugal and possessions
- France and possessions
- Netherlands and possessions
- Denmark and possessions
- Russia and possessions

> **Critical Viewing** In this painting, Columbus and his crew land in North America as Native Americans arrive to meet the explorers in their canoes. What qualities must explorers have had to undertake their voyages?

bravery, audacity, diplomacy, arrogance

Establishing Colonies

In addition to trade, Europeans used the voyages of exploration to claim lands for their own countries. When explorers landed in a new place, they declared it a colony. A **colony** is an area controlled by a distant country. As you have learned, Spanish explorers claimed colonies in Mexico and South America. The French and the English also established colonies in North America. By 1650, European countries controlled parts of Africa and Asia as well.

European exploration and colonization resulted in a sharing of goods and ideas known as the Columbian Exchange. From the Americas, Europeans obtained new foods, such as potatoes, corn, and tomatoes. Europeans introduced wheat and barley to the Americas. They also introduced diseases like smallpox. The diseases killed millions of native peoples.

Before You Move On
Monitor Comprehension What inspired Europeans to undertake voyages of exploration, and what did they gain as a result?

the desire to expand trade, spread Christianity, and gain colonies; they expanded trade and gained colonies as a result of the voyages

ONGOING ASSESSMENT
MAP LAB
GeoJournal

1. **Interpret Maps** According to the map, where in Asia did France establish a large colony? Why was this location beneficial geographically?

2. **Identify Problems and Solutions** Study the map. Who was Spain's main rival for colonies in South America? What problems might have arisen from their rivalry?

PLAN

OBJECTIVE Describe European voyages of exploration and the impact of colonization.

CRITICAL THINKING SKILLS FOR SECTION 3.1

- Main Idea
- Monitor Comprehension
- Interpret Maps
- Identify Problems and Solutions
- Analyze Causes
- Draw Conclusions
- Create Charts

PRINT RESOURCES

Teacher's Edition Resource Bank

- Reading and Note-Taking: Outline and Take Notes
- Vocabulary Practice: Words in Context
- **GeoActivity** Compare European Explorers

TECHTREK **myNGconnect.com**

Fast Forward!
Core Content Presentations
Teach *Exploration and Colonization*

Maps and Graphs
Online World Atlas: Early Colonization of Africa, Asia, and the Americas

Connect to NG
Research Links

Interactive Whiteboard
GeoActivity Compare European Explorers

Digital Library
NG Photo Gallery, Section 3

Also Check Out
GeoJournal in Student eEdition

BACKGROUND FOR THE TEACHER

Even after realizing that the Americas were continents, Europeans continued to search for a westward route to Asia. In 1497, John Cabot failed to find a northwest passage, a water route through North America. The Northwest Passage through Arctic waters was not found until 1906. In 1521, Ferdinand Magellan found a southern route through the Americas when he discovered what would come to be called the Strait of Magellan. However, the 1914 completion of the Panama Canal provided a more conveniently located route to Asia.

ESSENTIAL QUESTION

How did Europe develop and extend its influence around the world?

Section 3.1 describes how Europeans colonized many parts of the world, introducing European culture and ways of life to these other regions.

INTRODUCE & ENGAGE

Three-Step Interview Ask the class to think about places they would like to explore if they had the chance. Why do they want to go there? Have pairs of students take turns asking each other those questions. Then ask pairs to share their interview results with the class. **0:15** minutes

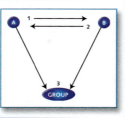

TEACH Maps and Graphs

Guided Discussion

1. Analyze Causes Discuss with students the dangers of exploration in the 1400s. **ASK:** What were European motives for exploration? *(Europeans wanted to find trades routes to Asia, to convert people in other lands to Christianity, and to claim other lands as colonies.)*

2. Draw Conclusions How do you think the introduction of foods from the Americas affected European life? *(Possible responses: Europeans had more variety of things to eat. Farmers could grow different crops, which might have helped them prosper.)*

Interpret Maps Help students interpret the Early Colonization of Africa, Asia, and the Americas map using the textbook or the **Online World Atlas**. Have a volunteer explain the relationship between the colors of European countries and the colors that appear in other regions. *(Colonies around the world are shaded with the same color as the European country that controlled them.)* **ASK:** What regions of North America and South America remained unclaimed? *(the northwest part of North America and the southern tip of South America)* Why do you suppose these areas remained unclaimed? *(They are the farthest from Europe, so explorers probably did not reach them by the date on this map.)* **0:15** minutes

DIFFERENTIATE Connect to NG

Striving Readers Create Charts Provide students with a three-column chart with the heads *Explorer, Sponsoring Country,* and *Accomplishment* similar to the one shown below. Have them work in pairs to fill out the chart with information from the lesson. If they need help getting started, give them the names of the explorers: Bartolomeu Dias, Vasco da Gama, Christopher Columbus, Jacques Cartier, and Sir Francis Drake.

EXPLORER	SPONSORING COUNTRY	ACCOMPLISHMENT

Gifted & Talented Research Caravels Have students use the **Research Links** to learn about caravels, the fast maneuverable ships developed by the Portuguese. Suggest that they make two drawings of the ship: an exterior side view and a cross section. Each drawing should have significant features labeled. Display student drawings so the rest of the class can learn about the ships.

ACTIVE OPTIONS

Interactive Whiteboard
GeoActivity

Compare European Explorers Have students work in pairs to complete the chart about the explorers and answer the critical thinking questions. **0:15** minutes

EXTENSION Have groups do additional research and write a profile about one of the explorers, making sure to explain why the explorer should be remembered. **0:15** minutes

NG Photo Gallery

Analyze Visuals Have students view photos of Columbus's ships from the **NG Photo Gallery**. Discuss the ships and ask students to imagine traveling on them for long periods of time. **ASK:** What kinds of food do you think the explorers brought onto the ships? *(food that didn't spoil quickly)* What do you think the living conditions were like? *(very rough and spartan)* Why might cleanliness have been a problem? *(They probably had only minimal facilities for washing up.)* **0:15** minutes

On Your Feet

Explorer Jigsaw Divide the class into four groups. Assign each group an explorer: Dias, da Gama, Columbus, or Cartier. Give the groups five minutes to review the lesson and create a simple map that roughly shows the explorer's route based on the description in the book. Then have the members of each group count off by using letters of the alphabet. Rearrange students using a jigsaw, with all the A's meeting together and so forth. In the new groups, students should use their combined knowledge to create one map of all the explorers' routes. **0:20** minutes

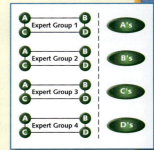

ONGOING ASSESSMENT
MAP LAB GeoJournal

ANSWERS

1. India; India was a source of many goods for trade and was therefore important.
2. Spain's main rival was Portugal. Their rivalry probably caused each to try to compete for territory. It may have led to violence on the sea, in South America, and at home.

3.2 The Industrial Revolution

TECHTREK
myNGconnect.com For an online map and images of the Industrial Revolution

Maps and Graphs | Digital Library

Main Idea The Industrial Revolution was an age of great developments in technology that changed how people worked and lived.

The Age of Exploration opened up trade around the world and brought great wealth to many western European countries. To increase this wealth, businesses looked for new ways to expand production. The result was the **Industrial Revolution**, a period when industry grew rapidly, and the production of machine-made goods greatly increased.

The Revolution Begins

The Industrial Revolution started in Great Britain in the 1700s as a result of new inventions and technologies. The **textile** industry, which deals with the manufacturing of cloth, was the first to be transformed by the revolution. In 1769, textile manufacturers began using machines that were run using water from a stream. Then, around 1770, James Hargreaves invented the spinning jenny. This machine allowed workers to make cotton and wool yarn at a much faster rate.

Before these inventions, most people made cloth by hand in their homes. However, the new machines were too large and expensive to use in small houses. Instead, the machines were placed in factories, and workers manufactured the goods there. In these early factories, each person worked on a small part of the product. This way of producing goods is called the **factory system**.

At first, factories were powered by water. Then around 1776, James Watt developed the steam engine, which was powered by coal. As a result, coal became an important raw material, and Britain benefited from its rich deposits of the fuel.

In the late 1700s, the Industrial Revolution spread to the rest of Europe. France and Belgium became leading manufacturers of textiles. Germany built factories for processing iron. Railroad systems developed in the 1800s. In 1825, George Stephenson built the first railroad in England. By 1850, thousands of miles of tracks crossed Europe.

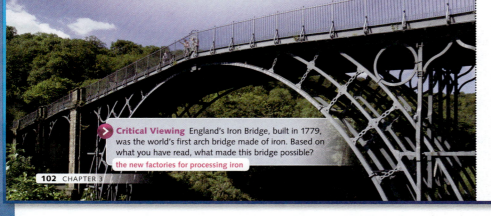

> **Critical Viewing** England's Iron Bridge, built in 1779, was the world's first arch bridge made of iron. Based on what you have read, what made this bridge possible?
> the new factories for processing iron

INDUSTRIES IN EUROPE, 1840–1890

Coal
Iron ore
Textiles
Railroad
International boundary

Impact of the Revolution

The Industrial Revolution had a tremendous impact on how people worked and lived. Cities grew rapidly because people migrated there for factory jobs. Standards of living rose, and a prosperous middle class grew.

However, factory workers often faced harsh conditions. Laborers worked as many as 16 hours a day. Child labor was common. Boys and girls as young as five years of age worked in factories and mines. Some were chained to their machines.

Many workers lived in small, crowded houses in neighborhoods where open sewers were common. Diseases spread quickly in these cramped buildings.

Over time, the workers' quality of life improved as sewer systems were created and other public health acts were passed.

Before You Move On
Summarize In what ways did the Industrial Revolution change how people lived and worked?
More people worked in factories, and cities grew rapidly.

ONGOING ASSESSMENT
MAP LAB
GeoJournal

1. **Interpret Maps** Where in Europe were most of the industries concentrated? What does this suggest about the economies of countries in other parts of Europe?

2. **Human-Environment Interaction** Which industry was the most widespread in Europe? Why was this such an important industry?

3. **Evaluate** Based on the map, which countries probably imported the fewest raw materials?

PLAN

OBJECTIVE Analyze and evaluate how industrialization changed European economies and people's way of life.

CRITICAL THINKING SKILLS FOR SECTION 3.2

- Main Idea
- Summarize
- Interpret Maps
- Evaluate
- Analyze Visuals
- Analyze Effects

PRINT RESOURCES

Teacher's Edition Resource Bank

- Reading and Note-Taking: Form and Support Opinions
- Vocabulary Practice: Cause and Effect Paragraph
- **GeoActivity** Evaluate Industrial Revolution Inventions

TECHTREK myNGconnect.com

>> **Fast Forward!**
Core Content Presentations
Teach *The Industrial Revolution*

Maps and Graphs
Interactive Map Tool
Analyze the Impact of Industrialization

Digital Library
NG Photo Gallery, Section 3

Also Check Out
- Online World Atlas in **Maps and Graphs**
- GeoJournal in **Student eEdition**

Connect to NG
Research Links

BACKGROUND FOR THE TEACHER

One development that helped lead to the Industrial Revolution was an agricultural revolution that took place during the 18th and 19th centuries. Changes such as new technology, better drainage, scientific animal breeding, new crops, and crop rotation all helped increase food production. Having bigger food supplies made it possible for societies to support a larger nonagricultural population, which freed people to work in manufacturing.

ESSENTIAL QUESTION

How did Europe develop and extend its influence around the world?

Section 3.2 discusses the benefits and drawbacks of the Industrial Revolution and the ways in which it changed how people lived and worked.

INTRODUCE & ENGAGE
Digital Library

Analyze Visuals Project the photos of a steam engine, textile factory, and railroad from the **NG Photo Gallery**. After showing the photos, **ASK:**

1. What advantage would steam power have over machines powered by animals or human energy?
2. How might the faster production of cloth change daily life?
3. How would a faster form of transportation help other industries?

Hands-On Geography Divide the class into two groups, one twice as large as the other. Tell the larger group to move their desks close together at one side of the room. Tell the smaller group to spread out their desks slightly. Move the class trash can to the middle of the spread-out group, and place a water bottle on one of their desks. Tell the class to suppose that these groups represent two neighborhoods. In one, people have plenty of space, sewers, water, and trash collectors. In the other, people live in crowded buildings without sewers or water. **ASK:** Which neighborhood will have more health problems? Explain. `0:20` **minutes**

TEACH

Guided Discussion

1. **Summarize** Discuss the Industrial Revolution. **ASK:** How did the Industrial Revolution change work life? *(Before the Industrial Revolution, most people worked in homes or on farms. Afterward, many people worked in factories for long hours under harsh conditions.)*

2. **Analyze Effects** How did the Industrial Revolution change transportation? *(Railroads were built, connecting many places in Europe.)*

MORE INFORMATION

Cottage Industries Before industrialization, much cloth was manufactured at home. Rural peasants could not farm during the entire year, so they spent some of their time spinning wool and weaving cloth in their cottages. Urban entrepreneurs began to organize these efforts by purchasing raw materials, distributing them to working families, and then marketing the finished products. Once factories became established, this system was abandoned.

Early machinery

DIFFERENTIATE
Connect to NG

English Language Learners **Give a Thumbs Up or Thumbs Down** Write a set of true-false statements about the lesson, such as "The textile industry was the first to industrialize." Read the lesson aloud with students following along in their books. Then have them close the books and listen as you read the true-false statements. Students should give a thumbs up if a statement is true and a thumbs down if a statement is false.

Pre-AP **Research Industrial Inventors** Have students use the **Research Links** to research important inventors of the Industrial Revolution. Ask students to choose one and write a paragraph about the person's life, using a drawing or photograph to illustrate it. Post the paragraphs in class on an inventors' wall or bulletin board.

ACTIVE OPTIONS

Interactive Map Tool

Analyze the Impact of Industrialization

PURPOSE Analyze how industrialization affects ecosystems

SET-UP

1. Open the **Interactive Map Tool**, set the "Region" to Europe, and set the "Map Mode" to Terrain.
2. Under "Environment and Society," turn on the Land Cover layer. Set the transparency level to 0 percent.

ACTIVITY

1. Point out the legend. **ASK:** Which of the categories most likely represents industrialized areas? *(Urban and Built-up)*
2. Then turn on the Human Footprint layer. Slide the transparency level to about 0 percent. Remind students that this map shows how human activity has affected ecosystems.
3. Slide the transparency level for both Land Cover and Human Footprint to about 40 percent. **ASK:** What inference can you make about the effect industrialization has had on ecosystems? *(The places that are most built up have the largest human footprint. From this, one can infer that industrialization changes ecosystems.)* `0:15` **minutes**

On Your Feet

Defend a Viewpoint On one wall of your classroom, post a sign that says, "Industrialization had mostly positive effects." On the opposite wall, post a sign that says, "Industrialization had mostly negative effects." Tell students to find the position on a line between the two signs that best represents their opinion. Have students explain their viewpoint. `0:20` **minutes**

ONGOING ASSESSMENT

MAP LAB
GeoJournal

ANSWERS

1. Great Britain; their economies probably weren't doing as well
2. Coal mining was widespread. Coal was used to fire machines such as steam engines.
3. Great Britain and the German Empire

TECHTREK
myNGconnect.com For images
of the French Revolution
Digital Library

3.3 The French Revolution

Main Idea The late 1700s in France was a period of economic and political unrest, which led to the French Revolution and the rise of Napoleon.

By the summer of 1789, the French people had not yet benefited from the Industrial Revolution. Harvests were poor, and prices skyrocketed. On July 14, mobs attacked the Bastille, Paris's ancient prison. This action sparked the French Revolution.

Roots of the Revolution

For years, France's lower and middle classes had suffered injustices. French society was composed of three large groups, called the Three Estates. The First Estate was made up of clergy. The Second Estate was made up of the nobility, or aristocrats. The Third Estate included everyone else, from merchants to peasants. The Third Estate paid most of the taxes but had no voice in government.

The people of the Third Estate began to call for change. Many of them were influenced by the **Enlightenment**. This movement stressed the rights of the individual. The ideas of Enlightenment thinkers like Voltaire and **John Locke** had helped inspire the American Revolution in 1776. The American Revolution, in part, inspired the revolution in France.

The Revolution Begins

In May 1789, the Third Estate demanded reforms, but the king of France, Louis XVI, refused. In response, the Third Estate formed the National Assembly. On August 26, 1789, the assembly issued the *Declaration of the Rights of Man and of the Citizen*. This document guaranteed liberty, equality, and property to citizens. The assembly tried to form a new government in which Louis would share power with an elected legislature. However, he again refused to cooperate.

The Radicals Take Over

Finally, in 1792, the Jacobins, a group of **radicals**, or extremists, seized power and formed the National Convention. The following year, the group executed Louis XVI and Marie Antoinette, his queen.

The violence soon got worse. Jacobin leader Maximilien Robespierre led a **Reign of Terror**. The Jacobins used a machine called the **guillotine** (GHEE uh teen) to cut off the heads of an estimated 40,000 people. In July 1794, the French finally turned on Robespierre and executed him.

Napoleon's Rise

After five years of violence, the French were exhausted. France was at war with Prussia, Austria, and Britain, and the government was not ruling effectively.

A young general, **Napoleon Bonaparte**, saw his chance and overthrew the government. Over the next five years, Napoleon increased his powers. He then declared himself Emperor Napoleon I and set about conquering other European powers and building an empire. Britain and Prussia finally defeated him in 1815.

Before You Move On
Summarize What led to the French Revolution and the rise of Napoleon?

high food prices, high taxes on the Third Estate, lack of a voice in government, ineffective rulers

△ Critical Viewing Marie Antoinette, shown here, was often accused of reckless spending. What details in this painting support this accusation?

her expensive clothing, jewelry, and elaborate hairstyle

ONGOING ASSESSMENT
SPEAKING LAB
GeoJournal

Express Ideas Through Speech Get together in a group and do research to prepare a panel discussion in which you will present the viewpoints of various figures from this section.

Step 1 Decide who each person in your group will be. You might choose from King Louis XVI, Marie Antoinette, Maximilien Robespierre, and Napoleon or be a member of the Third Estate.

Step 2 Come up with a few questions that your panel will discuss. The questions should focus on the French Revolution, the Reign of Terror, and Napoleon's rise.

Step 3 Present the panel discussion before the class. At its conclusion, invite questions and answer them in character.

French citizens stormed the Bastille because they thought it held guns and gunpowder.

The guillotine was considered an efficient and painless method of execution.

Statue of Napoleon on horseback

1785 — **1790** — **1795** — **1800** — **1805** — **1810**

1793 King Louis XVI and Marie Antoinette are executed; Reign of Terror begins.

1804 Napoleon names himself Emperor.

1789 Mobs attack the Bastille.

1792 Jacobins seize power.

1794 Robespierre is executed, and the Reign of Terror ends.

1799 Napoleon overthrows the French government.

1815 Napoleon is defeated.

PLAN

OBJECTIVE Summarize the causes and effects of the French Revolution and Napoleon's rise.

CRITICAL THINKING SKILLS FOR SECTION 3.3

- Main Idea
- Summarize
- Express Ideas Through Speech
- Draw Conclusions
- Form and Support Opinions
- Interpret Time Lines

PRINT RESOURCES

Teacher's Edition Resource Bank

- Reading and Note-Taking: Analyze Cause and Effect
- Vocabulary Practice: I Read, I Know, and So
- **GeoActivity** Map Napoleon's Empire

TECHTREK myNGconnect.com

▶▶ Fast Forward!
Core Content Presentations
Teach *The French Revolution*

Digital Library
NG Photo Gallery, Section 3

Also Check Out
- Graphic Organizers in **Teacher Resources**
- GeoJournal in **Student eEdition**

BACKGROUND FOR THE TEACHER

Louis XVI contributed to the revolutionary forces that caused his downfall. He gave both financial and military support to the American colonists in their revolution against Great Britain. To pay for France's involvement in the war, the government had to borrow money. Thus, Louis' support for the American Revolution contributed to the financial problems of France and so added to his unhappy subjects' concern.

ESSENTIAL QUESTION

How did Europe develop and extend its influence around the world?

European monarchies feared the spread of revolution and went to war against France, but as discussed in Section 3.3, the ideas of the Enlightenment thinkers and the belief that people could rebel continued to spread.

INTRODUCE & ENGAGE

Three-Step Interview Have students form pairs. Tell them to suppose that they are peasants who live in France just before the revolution. Have them take turns asking each other the questions: "How do you feel about the wealth of the royal family? What changes would you like to see in France?" After each partner has answered, have pairs form groups of four and share their interview results within the group. Correct any misconceptions. `0:15` minutes

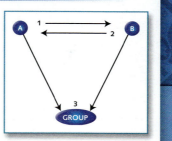

TEACH

Guided Discussion

1. Draw Conclusions Why did the American Revolution help lead to the French Revolution? *(Possible response: The American Revolution demonstrated that people could overthrow a monarchy and establish a democratic government based on Enlightenment ideas.)*

2. Form and Support Opinions How do you think the French Revolution might have been different if Louis XVI had agreed to make reforms? Why? *(Possible responses: The king might not have been executed if he had agreed to share power. If the king had cooperated, conditions might have improved so the radicals would not have seized power.)*

Interpret Time Lines Direct students' attention to the time line in the lesson. **ASK:** Where would you add the following event: the publication of the *Declaration of the Rights of Man and of the Citizen*? *(1789)* How much time passed between the start of the revolution and Napoleon's overthrow of the revolutionary government? *(10 years)* How much time passed from Napoleon's coronation as emperor to his final defeat? *(11 years)* `0:15` minutes

DIFFERENTIATE

Striving Readers **Analyze Effects** Give students a two-column Cause-and-Effect chart on the French Revolution with the causes already listed in the left-hand column, as shown below. Ask students to work with a partner to find the effects of each cause and record them on the chart.

Causes	Effects
• Bread prices skyrocketed. • Louis XVI refused to make reforms. • Radicals seized the French government. • The Jacobins executed 40,000 people. • While at war, the government did not rule effectively.	

Gifted & Talented **Build a Social Network** Have students choose one person from the French Revolution, such as Louis XVI, Marie Antionette, Robespierre, Napoleon, or someone else they learn about through research. Tell them to design this person's page on a social network site. The page can include photos, quotations, posts about daily activities, and lists of historical figures who decided to "friend" or "unfriend" that person. Allow students to display their pages in class.

NG Photo Gallery

Analyze Visuals Have students view images of the French Revolution from the **NG Photo Gallery** and answer the following questions with a partner.

GUILLOTINE
Why do you think crowds gathered to watch the executions?

NAPOLEON
What do you think his posture and facial expression are meant to convey?

ROBESPIERRE
Judging from this image, how would you describe Robespierre's character?

THE BASTILLE
What impression do you get of the Bastille from this image?

MARIE ANTIONETTE
How would you describe the expression on Marie Antionette's face?

LOUIS XVI
How does the image of Louis XVI compare to other paintings of rulers you have seen? `0:25` minutes

On Your Feet

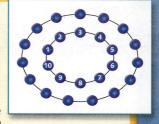

Fishbowl Post the question *Was the French Revolution a success or a failure?* Explain that to answer this question, students must first determine the revolutionaries' goals and then evaluate whether those goals were met. Have one-half of the class sit in a close circle, facing inward. The other half of the class sits in a larger circle around them. Students in the inner circle should discuss the question for ten minutes while those in the outer circle listen to the discussion and evaluate the points made. Then have the groups reverse roles and continue the discussion. `0:20` minutes

ONGOING ASSESSMENT

SPEAKING LAB GeoJournal

ANSWERS
The panel discussion should focus on the French Revolution, the Reign of Terror, and Napoleon. Students on the panel should present the viewpoint of the historical figures they are representing.

3.4 Declarations of Rights

TECHTREK
myNGconnect.com For photos of the documents and Guided Writing

Digital Library   Student Resources

MEN ARE BORN AND REMAIN FREE AND EQUAL IN RIGHTS.
— DECLARATION OF THE RIGHTS OF MAN AND OF THE CITIZEN

As you have learned, thinkers like John Locke and Voltaire led the Enlightenment. They asserted that people have ==natural rights==, or rights that people possess at birth, such as life, liberty, and property. Two key documents describe these rights: the American Declaration of Independence and the French *Declaration of the Rights of Man and of the Citizen*. In 1993, Nelson Mandela of South Africa received the Nobel Peace Prize. In his speech at the ceremony, he explained that the rights detailed in the declarations are still important.

Mandela and fellow Nobel recipient, F. W. de Klerk, were elected co-presidents of South Africa in 1994.

DOCUMENT 1

from the **Declaration of Independence** (July 4, 1776)

We hold these truths to be self-evident, that all men are created equal, that they are endowed [provided] by their Creator with certain unalienable [guaranteed] Rights, that among these are Life, Liberty, and the pursuit of Happiness; that, to secure these rights, Governments are instituted among Men, deriving their just powers from the consent of the governed.

This painting illustrates the signing of the Declaration of Independence.

CONSTRUCTED RESPONSE

1. What rights are citizens guaranteed?

DOCUMENT 2

from the **Declaration of the Rights of Man and of the Citizen** (August 26, 1789)

The representatives of the French people, organized as a National Assembly, . . . have determined to set forth in a solemn declaration the natural, unalienable, and sacred rights of man. Articles:

1. Men are born and remain free and equal in rights. Social distinctions [classes] may be founded only upon the general good.

2. The aim of all political association is the preservation of the natural . . . rights of man. These rights are liberty, property, security, and resistance to oppression.

CONSTRUCTED RESPONSE

2. Think about what you learned in Section 3.3 about the roots of the French Revolution. In what ways might the ideas in this document have inspired the French people to revolt?

DOCUMENT 3

from **Nobel Lecture** by Nelson Mandela (December 10, 1993)

Nelson Mandela helped lead the struggle to end ==apartheid== (uh PAHRT hyt) in South Africa. This system had denied black South Africans their rights. In recognition of his efforts, Mandela received the Nobel Peace Prize. The following excerpt is from his acceptance speech.

The value of our shared reward will and must be measured by the joyful peace which will triumph, because [of] the humanity that bonds both black and white into one human race. . . .

Thus shall we live, because we will have created a society which recognizes that all people are born equal, with each entitled in equal measure to life, liberty, prosperity, human rights, and good governance.

CONSTRUCTED RESPONSE

3. How do the rights Mandela discusses reflect those described in Documents 1 and 2?

ONGOING ASSESSMENT
WRITING LAB
GeoJournal

DBQ Practice Think about the ideas in the Declaration of Independence and the *Declaration of the Rights of Man and of the Citizen*. How did these ideas influence Nelson Mandela?

Step 1. Review your answers to Constructed Response questions 1, 2, and 3.

Step 2. On your own paper, jot down notes about the main ideas expressed in each document.

Document 1: Declaration of Independence

Main Idea(s) _____

Document 2: Declaration of the Rights of Man and of the Citizen

Main Idea(s) _____

Document 3: Nobel Lecture

Main Idea(s) _____

Step 3. Use your notes to construct a topic sentence that answers this question: How did the Declaration of Independence and the *Declaration of the Rights of Man and of the Citizen* influence Nelson Mandela?

Step 4. Write a paragraph that explains specific phrases and ideas in the Declaration of Independence and the *Declaration of the Rights of Man and of the Citizen*. Go to **Student Resources** for Guided Writing support.

PLAN

OBJECTIVE Analyze the philosophical ideas about human rights on which democracy is based.

CRITICAL THINKING SKILLS FOR SECTION 3.4

- Explain
- Compare and Contrast

PRINT RESOURCES

Teacher's Edition Resource Bank

- Reading and Note-Taking: Analyze Primary Sources
- Vocabulary Practice: Meaning Map
- **GeoActivity** Analyze Primary Sources: Women's Rights

TECHTREK myNGconnect.com

⏩ **Fast Forward!**
Core Content Presentations
Teach *Declarations of Rights*

Teacher Resources
Graphic Organizers

Connect to NG
Research Links

Interactive Whiteboard
GeoActivity Analyze Primary Sources: Women's Rights

Also Check Out
- NG Photo Gallery in **Digital Library**
- Writing Templates in **Teacher Resources**
- GeoJournal in **Student eEdition**

BACKGROUND FOR THE TEACHER

Another document closely related to these was the *Declaration of the Rights of Woman and of the Female Citizen* by Olympe de Gouges written in 1791. De Gouges strongly advocated free speech for women and acted on this belief by denouncing the Jacobins for killing the king. For this criticism, she was put to death on the guillotine.

ESSENTIAL QUESTION

How did Europe develop and extend its influence around the world?

Section 3.4 explores the ideas of European Enlightenment philosophers, which helped spark the American and French revolutions. These ideas continue to influence the ideals held around the world about individual rights and democracy.

INTRODUCE & ENGAGE

Prepare for the Document-Based Question Introduce the class to the topic of human rights by having groups fill out a Venn diagram that compares and contrasts the rights that young people have with the rights that adult citizens have. Help students get started by asking the following questions:

Venn Diagram

- What rights do adults have that are denied to young people?
- Do you think this is fair?
- What further rights should be given to young people? Why? `0:15` minutes

TEACH　　Teacher Resources

Guided Discussion

1. **Explain** Read aloud and then discuss the excerpt from the *Declaration of the Rights of Man and of the Citizen.* **ASK:** According to the *Declaration of the Rights of Man and of the Citizen,* what is the only reason to have social classes? *(to achieve the general good)*

2. **Contrast** What differences do you notice between Mandela's speech and the two earlier declarations? *(Possible responses: He spoke of joyful peace, which was not mentioned by the earlier documents. He refers to humanity and people, not men, which means he wants to make sure women are included. He mentions good governance as a human right.)*

Compare and Contrast Find and copy the Y-Chart in the **Graphic Organizers** and distribute it to students. Have them fill out the chart to analyze the ways in which the Declaration of Independence and the *Declaration of the Rights of Man and of the Citizen* are similar and different. `0:20` minutes

DIFFERENTIATE　　Connect to NG

Inclusion Synthesize Help students minimize distractions by typing the three excerpts on one sheet of paper. Give photocopies of these to students along with highlighters. Tell students to highlight important words that appear in all three documents. Then have them write a summary sentence using several of the words.

Pre-AP Trace the Roots of Democracy Have students use the **Research Links** to learn what previous documents influenced Thomas Jefferson as he wrote the Declaration of Independence. They can research the Magna Carta, the English Bill of Rights, or the writings of John Locke. Use a chart like the one below to take notes. Then have them write an expository paragraph describing how the source they studied influenced Jefferson.

MAGNA CARTA	ENGLISH BILL OF RIGHTS	JOHN LOCKE

ACTIVE OPTIONS

Interactive Whiteboard
GeoActivity

Analyze Primary Sources: Women's Rights Before students analyze the excerpts from *A Vindication of the Rights of Woman* and the letter from Abigail Adams, have pairs discuss women's rights today. Encourage them to consider what, if any, inequalities still exist between men and women. After they have read the excerpts, ask students whether they believe Abigail Adams and Mary Wollstonecraft's hopes for women have been met or exceeded. `0:20` minutes

On Your Feet

Human Rights Roundtable Divide the class into groups of four or five. Have the groups move desks together to form a table where they can all sit. Hand each group a sheet of paper with the question *What basic rights do you think all people should have?* The first student in each group should write an answer, read it aloud, and pass the paper clockwise to the next student. Each student in the group should add at least one answer. The paper should circulate around the table until students run out of answers or time is up. `0:15` minutes

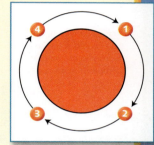

CONSTRUCTED RESPONSE ANSWERS

1. life, liberty, the pursuit of happiness
2. The ideas inspired them to demand their rights and overthrow the king.
3. He asserts that all people are born equal and that they are entitled to life, liberty, prosperity, human rights, and good governance.

ONGOING ASSESSMENT
WRITING LAB　　GeoJournal

ANSWERS

Steps 1 and 2 Responses will vary.
Step 3 The human rights guaranteed in the Declaration of Independence and the *Declaration of the Rights of Man and of the Citizen* influenced Nelson Mandela to demand these same rights for all of the people of South Africa.
Step 4 Students' paragraphs should have clear topic sentences and should synthesize ideas from the documents. Paragraphs should include references to equality and the rights of life and liberty.

3.5 Nationalism and World War I

TECHTREK
myNGconnect.com For an online map of Europe before World War I

Maps and Graphs

Main Idea Nationalism, new alliances, and growing tensions in Europe led to World War I.

After the French Revolution, the French people developed powerful feelings of nationalism. **Nationalism** is a strong sense of loyalty to one's country. During the 1800s, nationalism swept through Europe.

Italy and Germany Unify

Nationalism led to unification efforts in Italy and Germany. In 1800, the Italian Peninsula was made up of separate city-states. In 1870, the states came together to form a unified Italy. Germany was also composed of many different states in the early 1800s. Beginning in 1865, Prussia, the most powerful German state, led the way to unification. Driven by nationalist feelings, Prussia fought to take control of other German states away from their non-German rulers. In 1871, the states came together as a united German Empire.

Growing Tensions in Europe

By 1900, tensions had begun to grow among European powers. Nationalism had united some countries from within. However, nationalism also created fierce competition among rival countries.

Mainly, the countries competed for raw materials and colonies in Africa and Asia. To strengthen their position, Britain, France, and Russia formed an **alliance**, or agreement to work toward a common goal, called the Triple Entente. The German Empire and Austria-Hungary formed an alliance known as the Central Powers.

These alliances were tested in June 1914, when Archduke Franz Ferdinand of Austria-Hungary was assassinated in Serbia by a nationalist from Bosnia-Herzegovina. The assassin belonged to a group that was unhappy with Austrian rule of Bosnia-Herzegovina and wanted to unite with Serbia. Immediately after the assassination, Austria-Hungary declared war on Serbia. Then, because Serbia was a Russian ally, Russia declared war on Austria-Hungary. Within weeks, much of Europe had been drawn into war.

A Brutal War

The Great War, as it was called, dragged on for four brutal years. Both sides fought from **trenches**, or long ditches that protected soldiers from the enemy's gunfire. Both sides also used deadly technology, including machine guns, airplanes, tanks, and poison gas. German U-boats, or submarines, sank British ships.

In 1917, Germany seemed to gain an advantage when the Communist Party seized control of Russia's government and economy and made peace with Germany. That same year, the United States entered the war on the side of France and Britain. The fresh American troops helped turn the tide against Germany. In 1918, Germany surrendered to France, Britain, and the United States. By the time the war ended, ten million soldiers had died. About seven million civilians also lost their lives.

Impact of the War

In 1919, Germany signed the **Treaty of Versailles**. Under this peace treaty, Germany was forced to pay several billion dollars in damages and accept full blame for the war. Many of Germany's territories were taken away, and new countries were formed, including Austria, Hungary, Czechoslovakia, Yugoslavia, and Turkey. The treaty angered and humiliated the German people and did little to ease tensions in Europe. These tensions would help lead the way to another world war in a little more than 20 years.

EUROPE BEFORE WORLD WAR I 1914

Triple Entente	Neutral countries that joined the Central Powers
Neutral countries that joined the Triple Entente	Countries that remained neutral
Central Powers	

Nationalism created strong feelings of loyalty to one's country, and the new alliances pitted groups of countries against other groups. Tensions mounted because when a country belonging to one alliance was attacked, the entire alliance had agreed to retaliate.

Before You Move On

Monitor Comprehension In what ways did nationalism, new alliances, and growing tensions in Europe lead to World War I?

ONGOING ASSESSMENT

MAP LAB
GeoJournal

1. **Region** According to the map, which empires ruled much of Europe in 1914?

2. **Make Inferences** Note the countries that remained neutral in the war. What geographic factors might have encouraged their neutrality?

1870 Italy unifies.

Prussian prime minister Otto von Bismarck oversaw German unification.

Illustration of Archduke Ferdinand's assassination

Poster illustrating the alliance of Britain, France, and Russia in 1915

1917 United States enters the war.

1919 Treaty of Versailles is signed.

Leaders who signed the treaty

1870   **1885**   **1900**   **1915**   **1930**

1871 German states unite to form the German Empire.

1914 Archduke Ferdinand is assassinated; World War I begins.

1918 World War I ends.

French prime minister Georges Clemenceau

American president Woodrow Wilson

British prime minister David Lloyd George

PLAN

OBJECTIVE Explain the nationalist tensions and struggles for power that led to World War I.

CRITICAL THINKING SKILLS FOR SECTION 3.5

- Main Idea
- Monitor Comprehension
- Make Inferences
- Analyze Visuals
- Interpret Models
- Analyze Cause and Effect
- Evaluate
- Interpret Maps

PRINT RESOURCES

Teacher's Edition Resource Bank

- Reading and Note-Taking: Analyze Cause and Effect
- Vocabulary Practice: Definition Clues
- **GeoActivity** Analyze Causes and Effects of World War I

TECHTREK myNGconnect.com

Fast Forward!
Core Content Presentations
Teach *Nationalism and World War I*

Digital Library
NG Photo Gallery, Section 3

Maps and Graphs
Online World Atlas: Europe Before World War I, 1914

Connect to NG
Research Links

Interactive Whiteboard
GeoActivity Analyze Causes and Effects of World War I

Also Check Out
- Graphic Organizers in **Teacher Resources**
- GeoJournal in **Student eEdition**

BACKGROUND FOR THE TEACHER

Slavic nationalism played a large part in the events that triggered World War I. Serbia and Russia were independent Slavic nations, but many other Slavs lived in Austria-Hungary or the Ottoman Empire. A secret Serbian society called the Black Hand had pledged to liberate those Serbs who lived under foreign rule. A member of the group carried out the assassination of Archduke Franz Ferdinand. When Austria declared war on Serbia, Russia felt compelled to join the fight in support of its Slavic ally.

ESSENTIAL QUESTION

How did Europe develop and extend its influence around the world?

Section 3.5 discusses the growing tensions in Europe that led to World War I and the war's impact on the region.

INTRODUCE & ENGAGE Digital Library

Analyze Visuals Project the images of World War I from the **NG Photo Gallery**. Have the class draw conclusions about how warfare in the early 20th century was similar to and different from warfare today.

Interpret Models Divide the class into groups. Post the drawing below where students can see it and have them copy it on their own paper. Explain that it represents two trenches with an open field between them. Ask them to brainstorm how soldiers could cross the open field and capture the enemy trench without using any weapons other than guns and hand grenades and without being killed along the way. Explain that this was a problem faced by the armies in World War I. `0:15` minutes

TEACH Maps and Graphs

Guided Discussion

1. Analyze Cause and Effect Discuss the meaning of the word *nationalism*. **ASK:** How did nationalism cause tension in Europe? (*Nations competed for raw materials and colonies in Africa and Asia.*)

2. Evaluate Did the European alliance system of the early 1900s achieve its goal? Explain. (*Possible response: No. Countries form alliances for protection, but the alliance system of the early 1900s drew a whole continent into war.*)

Interpret Maps Have students review the material under the head "A Brutal War" and then study the map in the lesson or in the **Online World Atlas. ASK:** Which group of nations won the war, and which group of nations lost the war? (*The Triple Entente won, and the Central Powers lost.*) How did the map of Europe probably change after the end of the war? (*The alliance that lost the war had lands taken away to form independent nations.*) `0:10` minutes

DIFFERENTIATE Connect to NG

English Language Learners **Use Vocabulary Word Maps** Pair beginning and more advanced English Language Learners. Have them use a graphic organizer like the one below to make a Word Map for each of the three Key Vocabulary words from the lesson.

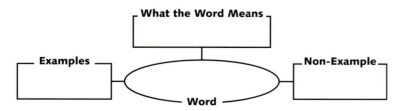

Gifted & Talented **Make a Poster** Suggest that students use the **Research Links** to find examples of World War I recruitment posters and then create their own. Remind students that recruitment posters are very nationalistic and that they offer persuasive reasons to fight for one particular country. Have students display their final posters in class.

ACTIVE OPTIONS

Interactive Whiteboard
GeoActivity

Analyze Causes and Effects of World War I Because of the complexity of this graphic organizer, start filling it out as a class. Ask for volunteers to provide the first two entries. Give the class time to record the answers on the chart. Then allow students to form small groups and work together to complete the Cause-and-Effect chart. Monitor groups' progress and correct any misconceptions. `0:15` minutes

EXTENSION Tell students that World War I changed the way war was fought. Ask students to use what they have learned and the photos they have seen about the war to identify how it changed warfare. Use the following to stimulate discussion:

- the different types of weapons used during the war
- the number of nations drawn into the conflict
- the length of the war
- the number of dead caused by it `0:10` minutes

On Your Feet

Think, Pair, Share Give students a few minutes to think about the question *Why was World War I such a deadly conflict?* Then have students choose partners and talk about the question for five minutes. Finally, allow individual students to share their ideas with the class. `0:15` minutes

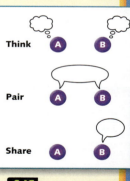

ONGOING ASSESSMENT
MAP LAB GeoJournal

ANSWERS
1. German Empire, Austria-Hungary, Russia, Ottoman Empire
2. Geographic factors included mountains and water barriers, such as the North Sea and Baltic Sea.

3.6 World War II and the Cold War

TECHTREK
myNGconnect.com For an
online map of post-war Europe

Maps and
Graphs

Main Idea After World War II was fought to defeat the Axis Powers, the Cold War developed between the democratic United States and the Communist Soviet Union.

At the end of World War I, Germany lost its military power. As you have read, the Treaty of Versailles also placed full blame for the war on Germany and forced it to pay **reparations**, or money to cover the losses suffered by the victors. The Great Depression, which began in 1929, further damaged Germany's economy. The **Great Depression** was a severe downturn in the world's economy. During this crisis, **Adolf Hitler** rose to power in Germany.

World War II

Hitler became the leader of the National Socialist German Workers' Party, or the Nazis. In 1936, Hitler made an alliance with Italy. Germany also formed an alliance with Japan, where the military had seized power. Germany, Italy, and Japan formed the Axis Powers.

Germany's invasion of Poland in 1939 started World War II. Two of Poland's allies, Great Britain and France, declared war on Germany soon after the invasion. Germany responded by conquering Poland and then quickly took over most of Europe, including France.

In 1941, Japan attacked the United States at Pearl Harbor, Hawaii. As a result, the United States abandoned its neutrality and entered the war on the side of Britain and the Soviet Union. Together, they were known as the Allies. Over time, many other countries took sides and joined either the Allies or the Axis Powers.

110 CHAPTER 3

After more than five years of war, Germany surrendered on May 8, 1945. Allied troops were stunned to find the Nazi **concentration camps** where six million Jews and other victims had been murdered. This mass slaughter was called the **Holocaust**. Japan continued to fight until the United States dropped atomic bombs on Hiroshima and Nagasaki. Japan surrendered on September 2, 1945.

The Cold War

After World War II, the Soviet Union established Communist governments in Eastern Europe. Germany was divided into Communist East Germany and democratic West Germany. The imaginary boundary that separated Eastern and Western Europe was called the **Iron Curtain**. The division marked the beginning of the **Cold War**, a period of great tension between the United States and the Soviet Union.

To defend against possible attack, both sides forged military alliances. Western Europe and the United States formed NATO (North Atlantic Treaty Organization), while Communist Eastern Europe formed the Warsaw Pact. The two never directly waged war against each other during the course of the Cold War.

In the 1980s, many eastern European countries overthrew their Communist governments. In 1991, the Soviet Union itself collapsed. The Cold War ended, and democracy replaced communism throughout Eastern Europe.

Before You Move On
Make Inferences In what ways did World War II help lead to the Cold War? After helping win the war, the Soviet Union emerged stronger than ever. Stalin used his increased power to begin taking over Eastern Europe, thereby setting up the philosophical divide of the Cold War.

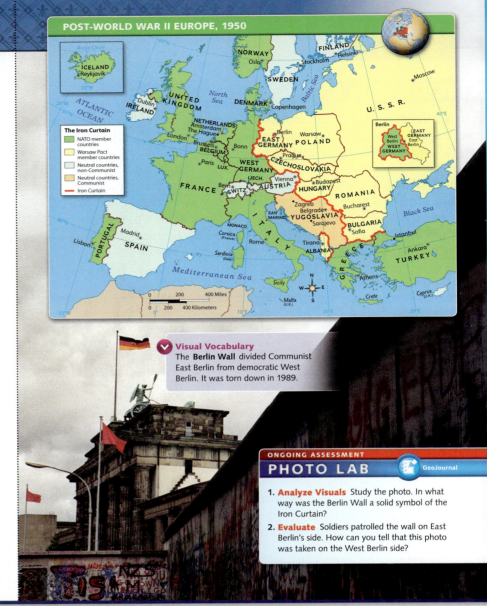

POST-WORLD WAR II EUROPE, 1950

The Iron Curtain
- NATO member countries
- Warsaw Pact member countries
- Neutral countries, non-Communist
- Neutral countries, Communist
- Iron Curtain

Visual Vocabulary
The **Berlin Wall** divided Communist East Berlin from democratic West Berlin. It was torn down in 1989.

ONGOING ASSESSMENT
PHOTO LAB
GeoJournal

1. **Analyze Visuals** Study the photo. In what way was the Berlin Wall a solid symbol of the Iron Curtain?

2. **Evaluate** Soldiers patrolled the wall on East Berlin's side. How can you tell that this photo was taken on the West Berlin side?

PLAN

OBJECTIVE Describe the conflicts that led to World War II and the Cold War.

CRITICAL THINKING SKILLS FOR SECTION 3.6

- Main Idea
- Make Inferences
- Analyze Visuals
- Evaluate
- Compare and Contrast
- Pose and Answer Questions

PRINT RESOURCES

Teacher's Edition Resource Bank

- Reading and Note-Taking: Sequence Events
- Vocabulary Practice: Word Map
- **GeoActivity** Analyze Results of World War II: Berlin

TECHTREK myNGconnect.com

Fast Forward!
Core Content Presentations
Teach *World War II and the Cold War*

Digital Library
GeoVideo: *Introduce Europe*

Connect to NG
Research Links

Maps and Graphs
Interactive Map Tool
Compare Europe Then and Now

Also Check Out
- NG Photo Gallery in **Digital Library**
- Online World Atlas in **Maps and Graphs**
- GeoJournal in **Student eEdition**

BACKGROUND FOR THE TEACHER

Germany learned from the deadly, drawn-out trench warfare of World War I. In World War II, the German army adopted a tactic known as *blitzkrieg*, or "lightning war." A blitzkrieg uses fast-moving troops, tanks, motorized artillery, and dive-bombers to attack targets whose destruction will undermine the enemy's ability to coordinate a defense. By using blitzkrieg tactics, Germany overran Poland, Belgium, the Netherlands, and France in a single year.

ESSENTIAL QUESTION

How did Europe develop and extend its influence around the world?

Section 3.6 discusses how Europe drew much of the world into World War II. The section also describes how many countries became involved in the Cold War.

INTRODUCE & ENGAGE 📼 Digital Library

GeoVideo: *Introduce Europe* Show students the portion of the video on World War II. Then **ASK:** How do you think the war affected Europe? *(killed many people, caused much destruction, freed many people)*

Hands-On Geography String a rope so that it divides the room in half. If possible, make sure that one side of the room has no access to exits. Tell students that a dictatorship has taken over the half without the exit and that no one from that side will be allowed to cross to the other side. In addition, the dictator will not allow students from the "free" side to visit or talk to students across the line. Ask students to identify the limits this arrangement places on them and their ability to do their schoolwork. Then introduce the idea that Europe was divided into two hostile regions after World War II. `0:10` minutes

TEACH 🖱 Connect to NG

Guided Discussion

1. **Make Inferences** Discuss the rise of Adolf Hitler. **ASK:** Why was a leader like Adolf Hitler able to take over Germany? *(Possible response: Germans were angry about their defeat in World War I and their economic struggles, so they might have thought Hitler would make the country strong again.)*

2. **Compare and Contrast** During the Cold War, what two types of governments were dominant in Western Europe and Eastern Europe? *(Democratic governments in Western Europe and Communist governments in Eastern Europe)*

Pose and Answer Questions Have groups of students prepare a list of questions they still have about World War II and the Cold War. To help get them started, provide a list of the 5Ws: *Who? What? Where? When?* and *Why?* Then have each student select a different question from the list and use the **Research Links** to find the answer. Encourage students to share the results of their research with the class. `0:20` minutes

DIFFERENTIATE 📱 Connect to NG

English Language Learners Explain Confusing Terms Clarify the potentially confusing term *Cold War*. First, explain that it was used to contrast the ongoing state of tension with "hot" wars in which shooting occurs. Then tell students that the word *cold* is often used to mean "unfriendly" or "hostile." Another example of this usage is the phrase *cold shoulder*.

Pre-AP Compare Across Regions Have students use the **Research Links** to create a chart that compares selected countries of Eastern and Western Europe. Have them choose four countries for each region and find out the latest GDP per capita for each country. Then have students use their findings to create a graph. Ask students to write a caption that draws a conclusion about the long-term effects of a free-market economy versus a command economy.

ACTIVE OPTIONS

Interactive Map Tool

Compare Europe Then and Now

PURPOSE Analyze how the nations of Europe have changed since the Cold War era

SET-UP
Open the **Interactive Map Tool**, set the "Region" to Europe and set the "Map Mode" to National Geographic. Close the toolbar to the left and adjust the map so that Europe fills as much of the screen as possible.

ACTIVITY
Have students open their books to the lesson and study the map of Post–World War II Europe. **ASK:** What nation, divided during the Cold War, has since been reunified? *(Germany)* What Eastern European nations today were once part of the Soviet Union? *(Estonia, Latvia, Lithuania, Belarus, Ukraine, and Moldova)* What happened to Yugoslavia and Czechoslovakia? *(Both broke apart into smaller countries.)* `0:15` minutes

On Your Feet

Living Time Line Divide the class into small groups. Each member of the group should choose one event from the lesson and write a brief description of it along with its date. Then have group members form a line so that their events are in order and read their descriptions aloud in sequence. `0:15` minutes

Performance Assessment

Express Ideas Through Speech Have students choose an event from Section 3 they feel strongly about and write and deliver a persuasive speech urging people to respond to it. For example, they could encourage people rise up against Louis XVI or they could urge Germans to protest the Berlin Wall. Remind them to use facts, data, and emotional appeals to persuade people to take action. Go to **myNGconnect.com** for the rubric.

ONGOING ASSESSMENT
PHOTO LAB 📓 GeoJournal

ANSWERS
1. Possible response: The Berlin Wall was almost literally an iron curtain that separated East Berlin from the free world.
2. The wall is covered with protest graffiti, which East German soldiers would have prevented.

VOCABULARY

Match each word in the first column with its definition in the second column.

WORD	DEFINITION
1. ecosystem	a. sold to relax penalty for sin
2. democracy	b. strong sense of loyalty to one's country
3. plebeians	c. community of living organisms and their environment
4. indulgence	d. area controlled by a distant country
5. colony	e. common people
6. nationalism	f. government of the people

MAIN IDEAS

7. What are some of the significant islands in Europe? (Section 1.2)

8. Where is much of Europe's farming industry located? (Section 1.3)

9. What events led to the development of democracy in ancient Greece? (Section 2.1)

10. How were plebeians represented in the Roman Republic? (Section 2.3)

11. In what way did the Roman Empire influence language? Why? (Section 2.4)

12. In what ways did the Roman Catholic Church serve as a unifying force in Western Europe? (Section 2.5)

13. What achievements in arts and literature did the Renaissance inspire? (Section 2.6)

14. How did the Industrial Revolution change Europe? (Section 3.2)

15. Why did the French people welcome Napoleon's rise to power? (Section 3.3)

16. Why did Russia declare war on Austria-Hungary in 1914? (Section 3.5)

17. What was the Cold War? (Section 3.6)

GEOGRAPHY

ANALYZE THE ESSENTIAL QUESTION

How did Europe's physical geography encourage interaction with other regions?

Critical Thinking: Evaluate

18. In what ways do rivers like the Danube make trade easier within Europe?

19. Why has trade been central to Europe's growth throughout its history?

EARLY HISTORY

ANALYZE THE ESSENTIAL QUESTION

How did European thought shape Western civilization?

Critical Thinking: Draw Conclusions

20. What elements of the democracy practiced in ancient Greece did the United States adopt?

INTERPRET MAPS

TRADE IN THE ROMAN EMPIRE

21. **Movement** From what part of the Empire did Rome obtain its grains? its textiles?

EMERGING EUROPE

ANALYZE THE ESSENTIAL QUESTION

How did Europe develop and extend its influence around the world?

Critical Thinking: Make Inferences

22. In what way did improvements in navigation and shipbuilding lead Europe to establish colonies in other parts of the world?

23. Why did the Industrial Revolution make European leaders eager to establish colonies in the Americas and Asia?

24. What are some of the positive effects of nationalism? What are some negative effects?

INTERPRET TABLES

MILES OF RAILWAY TRACK IN SELECTED EUROPEAN COUNTRIES (1840–1880)			
	1840	1860	1880
Austria-Hungary	144	4,543	18,507
Belgium	334	1,730	4,112
France	496	9,167	23,089
Germany	469	11,089	33,838
Great Britain	2,390	14,603	25,060
Italy	20	2,404	9,290
Netherlands	17	335	1,846
Spain	0	1,917	7,490

Source: Modern History Sourcebook

25. **Analyze Data** How much railway track did the Germans build between 1840 and 1880? What might account for this increase?

26. **Draw Conclusions** France and Spain are almost the same size. Note the difference in the extent of the railway system in each one in 1880. What does this suggest about the level of industrialization in each country?

ACTIVE OPTIONS

Synthesize the Essential Questions by completing the activities below.

27. **Write Tour Notes** Suppose that you are going to lead a group of tourists on a trip on one of Europe's rivers. Select and research the river. You might choose the Danube, Rhine, Tiber, Rhone, Thames, or Seine. Then write notes for a guided tour of the river. Start by describing its location, size, and appearance. Next, point out important sites along the river and explain their historical significance. Finally, discuss the river's uses today. **Gather photos of the river and conduct the tour with your group of "tourists."**

> **Writing Tips**
> - Use language that appeals to the senses to help your tourists see and experience the river.
> - Include stories about the sites and historical events to hold your audience's interest.
> - Involve your audience by comparing the river to one that is familiar to them.

TECHTREK myNGconnect.com For research links on European history

28. **Create a Slide Show** Prepare a slide show of famous European buildings, using **Connect to NG** or other online sources. Research and identify five buildings, such as the Pantheon in Rome, Italy. Write two sentences that explain the importance of each building to European history. Copy the chart below to help you organize your information.

BUILDING	IMPORTANCE TO EUROPE
1.	
2.	
3.	

CHAPTER Review

VOCABULARY ANSWERS

1. c
2. f
3. e
4. a
5. d
6. b

MAIN IDEAS ANSWERS

7. Great Britain, Ireland, Greenland, Iceland, Sicily, Corsica

8. on the Northern European Plain

9. Solon established assemblies in which wealthy people made laws. Cleisthenes established a direct democracy in which all citizens voted directly for laws.

10. by tribunes in an assembly

11. Latin, the language of Rome, spread throughout the empire and formed the basis of Romance languages.

12. The Church unified Western Europe around common religious beliefs.

13. many achievements in the arts and literature, including *The Divine Comedy* by Dante, *The Last Supper* by Leonardo da Vinci, *David* by Michelangelo

14. The Industrial Revolution introduced the factory system. People moved to cities to work in factories, and cities grew. Also, railroads were built across Europe.

15. They were tired of violence and ineffectual government. Napoleon brought order to France.

16. because Austria-Hungary declared war on Serbia, a country with which Russia was allied

17. a period of great tension between the United States and the Soviet Union

GEOGRAPHY

ANALYZE THE ESSENTIAL QUESTION ANSWERS

18. The Danube, like the Rhine, the Seine, and other rivers are navigable, and they connect different parts of Europe, allowing trade to occur among various regions.

19. because trade allowed European countries to interact with other lands, expand their wealth, and absorb new influences

EARLY HISTORY

ANALYZE THE ESSENTIAL QUESTION ANSWERS

20. The United States adopted the idea of having bodies of officials that represent the people and allowing people to vote for laws, although not directly.

INTERPRET MAPS

21. Grains came from eastern Europe, North Africa, and Egypt. Textiles came from Gaul, Italy, Egypt, and Syria.

EMERGING EUROPE

ANALYZE THE ESSENTIAL QUESTION ANSWERS

22. As Europeans developed improved techniques in navigation and shipbuilding, they were able to explore more parts of the world. As they explored, they established colonies.

23. They wanted raw materials to use in their new factories.

24. Responses will vary. Students' responses might include that nationalism helps build unity in a country to support common beliefs and values. Nationalism can be negative when people use it to exclude others or to try to expand into other countries' territories.

INTERPRET TABLES

25. Germany built 22,749 miles of track between 1840 and 1880. The demand for more railroads to transport goods and people might account for the increase.

26. The difference suggests that France was much more developed industrially.

ACTIVE OPTIONS

WRITE TOUR NOTES

27. Tour notes should
- have interesting introductions that describe the river;
- highlight at least three ways in which the river was historically important;
- emphasize the importance of the river to the country or countries in which it is located.

CREATE A SLIDE SHOW

For each photograph selected, students should clearly explain how the building has been important in Europe's history and development. See the chart at right for sample responses. →

BUILDING	IMPORTANCE TO EUROPE
1. Pantheon in Rome	The ancient temple contained the largest dome in the world for centuries. It is still considered an architectural marvel.
2. Parthenon in Athens	The temple established new standards for Greek architecture. It greatly influenced Western architecture.
3. Duomo of Florence	The cathedral became a symbol of the Italian Renaissance. It is still the largest masonry dome in the world.

CHAPTER PLANNER

SECTION SUPPORT

SECTION 1 CULTURE

1.1 Languages and Cultures

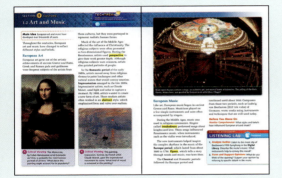

OBJECTIVE Draw conclusions about how countries preserve their languages and traditional cultures.

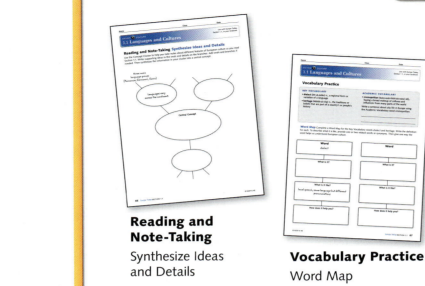

Reading and Note-Taking
Synthesize Ideas and Details

Vocabulary Practice
Word Map

Whiteboard Ready!

GeoActivity
Compare Urban Development

SECTION 1 CULTURE

1.2 Art and Music

OBJECTIVE Identify the styles of music and art associated with specific periods of European history.

Reading and Note-Taking
Sequence Events

Vocabulary Practice
Definition Clues

Whiteboard Ready!

GeoActivity
Recognize Architectural Movements

SECTION 1 CULTURE

1.3 Europe's Literary Heritage

OBJECTIVE Explain the relationships between literary movements and historical events.

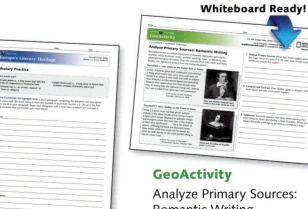

Reading and Note-Taking
Find Main Idea and Details

Vocabulary Practice
Compare/Contrast Paragraph

Whiteboard Ready!

GeoActivity
Analyze Primary Sources: Romantic Writing

ASSESSMENT

Student Edition
Ongoing Assessment: Photo Lab

Resource Bank and myNGconnect.com
Review and Assessment, Sections 1.1–1.4

ExamView®
Test Generator CD-ROM
Section 1 Quiz in English and Spanish

Student Edition
Ongoing Assessment: Listening Lab

Resource Bank and myNGconnect.com
Review and Assessment, Sections 1.1–1.4

ExamView®
Test Generator CD-ROM
Section 1 Quiz in English and Spanish

Student Edition
Ongoing Assessment: Writing Lab

Resource Bank and myNGconnect.com
Review and Assessment, Sections 1.1–1.4

ExamView®
Test Generator CD-ROM
Section 1 Quiz in English and Spanish

TECHTREK myNGconnect.com

 **Fast Forward!**
Core Content Presentations
Teach *Languages and Cultures*

Maps and Graphs
Interactive Map Tool
Analyze Language Diversity
of Europe

 Digital Library
NG Photo Gallery, Section 1

Also Check Out
GeoJournal in
Student eEdition

Connect to NG
Research Links

 **Fast Forward!**
Core Content Presentations
Teach *Art and Music*

Interactive Whiteboard
GeoActivity Recognize
Architectural Movements

 Digital Library
• GeoVideo: *Introduce Europe*
• NG Photo Gallery, Section 1
• Music Clip, Section 1

Also Check Out
GeoJournal in
Student eEdition

Connect to NG
Research Links

 Fast Forward!
Core Content Presentations
Teach *Europe's Literary Heritage*

Also Check Out
• NG Photo Gallery in
  Digital Library

Interactive Whiteboard
GeoActivity Analyze Primary
Sources: Romantic Writing

• Writing Templates in
  Teacher Resources
• GeoJournal in
  Student eEdition

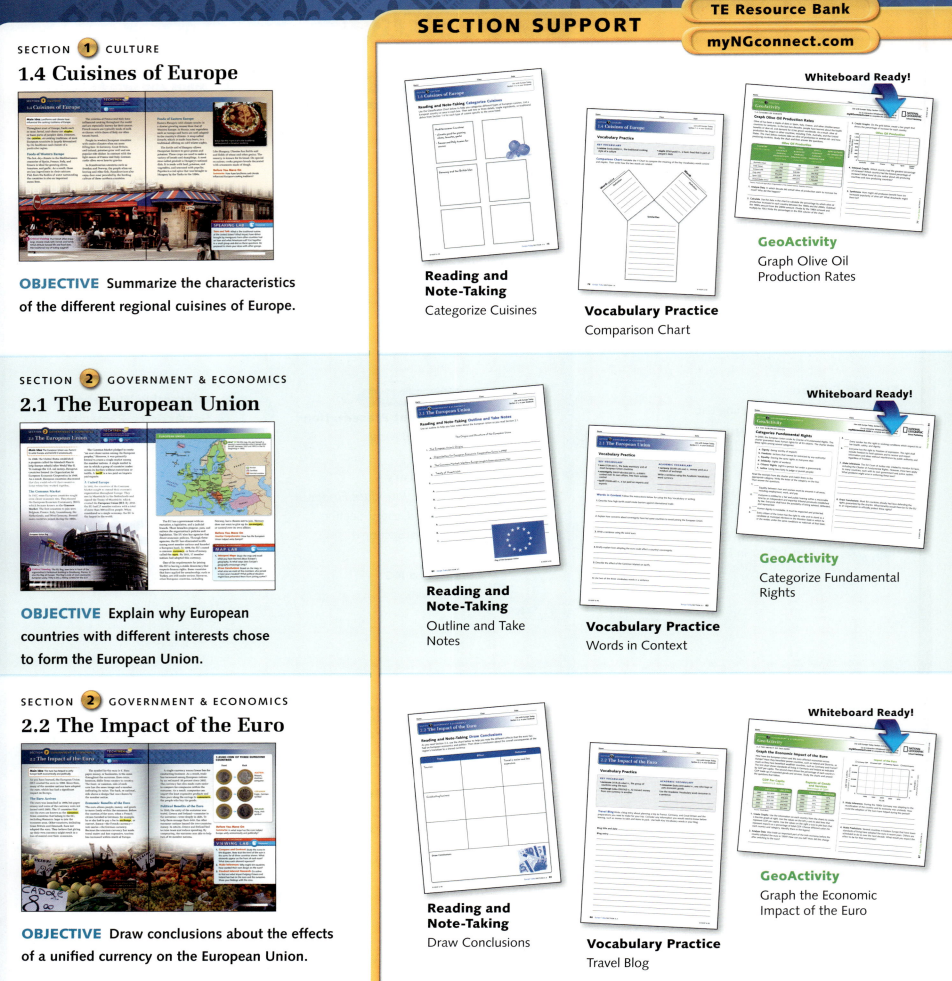

SECTION ① CULTURE

1.4 Cuisines of Europe

OBJECTIVE Summarize the characteristics of the different regional cuisines of Europe.

SECTION ② GOVERNMENT & ECONOMICS

2.1 The European Union

OBJECTIVE Explain why European countries with different interests chose to form the European Union.

SECTION ② GOVERNMENT & ECONOMICS

2.2 The Impact of the Euro

OBJECTIVE Draw conclusions about the effects of a unified currency on the European Union.

SECTION SUPPORT

TE Resource Bank

myNGconnect.com

Reading and Note-Taking
Categorize Cuisines

Vocabulary Practice
Comparison Chart

Whiteboard Ready!

GeoActivity
Graph Olive Oil Production Rates

Reading and Note-Taking
Outline and Take Notes

Vocabulary Practice
Words in Context

Whiteboard Ready!

GeoActivity
Categorize Fundamental Rights

Reading and Note-Taking
Draw Conclusions

Vocabulary Practice
Travel Blog

Whiteboard Ready!

GeoActivity
Graph the Economic Impact of the Euro

Student Edition
Ongoing Assessment: Speaking Lab

Teacher's Edition
Performance Assessment: Write Journal Entries

Resource Bank and myNGconnect.com
Review and Assessment, Sections 1.1–1.4

ExamView®
Test Generator CD-ROM
Section 1 Quiz in English and Spanish

Fast Forward!
Core Content Presentations
Teach *Cuisines of Europe*

Digital Library
NG Photo Gallery, Section 1

Interactive Whiteboard
GeoActivity Graph Olive Oil Production Rates

Also Check Out
- Graphic Organizers in **Teacher Resources**
- GeoJournal in **Student eEdition**

Student Edition
Ongoing Assessment: Map Lab

Resource Bank and myNGconnect.com
Review and Assessment, Sections 2.1–2.4

ExamView®
Test Generator CD-ROM
Section 2 Quiz in English and Spanish

Fast Forward!
Core Content Presentations
Teach *The European Union*

Digital Library
GeoVideo: *Introduce Europe*

Connect to NG
Research Links

Maps and Graphs
- **Interactive Map Tool** Explore EU Membership Patterns
- Online World Atlas: European Union

Also Check Out
- NG Photo Gallery in **Digital Library**
- GeoJournal in **Student eEdition**

Student Edition
Ongoing Assessment: Viewing Lab

Resource Bank and myNGconnect.com
Review and Assessment, Sections 2.1–2.4

ExamView®
Test Generator CD-ROM
Section 2 Quiz in English and Spanish

Fast Forward!
Core Content Presentations
Teach *The Impact of the Euro*

Teacher Resources
Graphic Organizers

Digital Library
NG Photo Gallery, Section 2

Interactive Whiteboard
GeoActivity Graph the Economic Impact of the Euro

Also Check Out
GeoJournal in **Student eEdition**

CHAPTER PLANNER

SECTION 2 GOVERNMENT & ECONOMICS

2.3 Democracy in Eastern Europe

OBJECTIVE Explain the challenges that eastern European countries face in making the transition from communism to democracy.

SECTION 2 GLOBAL ISSUES

2.4 Changing Demographics

OBJECTIVE Describe the impact of immigration on European people and culture.

CHAPTER ASSESSMENT

Review

SECTION SUPPORT

Reading and Note-Taking
Summarize Governments and Economies

Vocabulary Practice
Definition Chart

Whiteboard Ready!

GeoActivity
Research Reform Movements

Reading and Note-Taking
Analyze Cause and Effect

Vocabulary Practice
Related Idea Web

Whiteboard Ready!

GeoActivity
Compare London's Immigrant Populations

INFORMAL ASSESSMENT

Review and Assessment

Standardized Test Practice

Student Edition

Ongoing Assessment: Reading Lab

Resource Bank and myNGconnect.com

Review and Assessment, Sections 2.1–2.4

ExamView®
Test Generator CD-ROM

Section 2 Quiz in English and Spanish

Student Edition

Ongoing Assessment: Reading Lab

Teacher's Edition

Performance Assessment: Conduct a
Panel Discussion

Resource Bank and myNGconnect.com

Review and Assessment, Sections 2.1–2.4

ExamView®
Test Generator CD-ROM

Section 2 Quiz in English and Spanish

» Fast Forward!

Core Content Presentations

Teach *Democracy in Eastern Europe*

Digital Library

NG Photo Gallery, Section 2

Connect to NG

Research Links

Magazine Maker

Also Check Out

• Graphic Organizers in **Teacher Resources**

• GeoJournal in **Student eEdition**

» Fast Forward!

Core Content Presentations

Teach *Changing Demographics*

Maps and Graphs

Graph: Immigrant Populations in Europe, 2006

Interactive Whiteboard

GeoActivity Compare London's Immigrant Populations

Also Check Out

• Graphic Organizers in **Teacher Resources**

• GeoJournal in **Student eEdition**

Chapter Test A (on level)

Chapter Test B (modified)

ExamView®
Test Generator CD-ROM

Chapter Tests

STRATEGIES FOR DIFFERENTIATION

STRIVING READERS   e eEdition   ♪ Audiobook

Strategy 1 • Activate Prior Knowledge

Before reading, pose the following questions and have students brainstorm independently or in pairs to list ideas that come to mind. Call on volunteers to share list items as well as any additional information they have about their items.

- **1.1** How many European languages can you name?
- **1.2** Who are some famous European artists and musicians from the past?
- **1.3** What European authors or titles of famous works of literature can you name?
- **1.4** What do you know about foods from different European countries?

Use with Sections 1.1–1.4

Strategy 2 • Play "Who Am I?"

Create flip cards for the artists, musicians, and writers below, listing one identifying fact for each one. Use the cards to play a partner review game, with one partner reading a fact and the other partner naming the person associated with that fact.

Leonardo da Vinci	William Shakespeare
Claude Monet	Miguel de Cervantes
Ludwig van Beethoven	Jane Austen
Homer	Charles Dickens
Dante	James Joyce

Use with Sections 1.2 and 1.3 *Ask students to come up with an additional fact for each person.*

Strategy 3 • Record and Compare Facts

After reading a lesson, ask students to write two important facts they learned. Allow pairs of students to compare and check their facts and then combine their facts into one longer list. Ask a volunteer from each group to read the most important fact from the list.

Use with All Sections

Strategy 4 • Write Main Idea Questions

As they read, direct students to look for and make notes on what they think are key ideas that are important to remember. After reading, tell them to use their notes to write five questions about these key ideas. Allow students to exchange and answer each other's questions.

Use with Sections 2.1–2.4 *A main idea question for Section 2.1 might be "Why did the European Union form?"*

Strategy 5 • Expand Main Idea Statements

After reading, direct students to copy the Main Idea statement and write a paragraph that expands on the statement. Use the following starters as examples if needed:

- **2.2** The euro has helped to unify Europe both economically and politically. Some benefits of the euro include _____.
- **2.3** Eastern European countries have faced many challenges in their transition to democracy. For example, _____.
- **2.4** New immigrants are changing Europe. First, _____.

Use with Sections 2.2–2.4

INCLUSION   e eEdition   ♪ Audiobook

Strategy 1 • Describe Lesson Visuals

Pair visually impaired students with students who are not visually challenged. Ask the latter to help their partners "see" the visuals in the chapter by describing the images and answering any questions the visually impaired student might have.

Use with All Sections *For example, for the photograph in Section 1.1, students might describe the movement of the horses, the colorful costumes of the riders, and the excitement generated by the race.*

Strategy 2 • Preview the EU

Help students prepare for the lessons on the European Union by previewing the organization with the 5Ws:

Who: Many European countries
What: Formed a union
When: In 1992
Where: In Maastricht in the Netherlands
Why: To benefit economically

Use with Sections 2.1 and 2.2 *You can also use the 5Ws with Section 2.3 and focus the questions on one eastern European country.*

ENGLISH LANGUAGE LEARNERS 🔵 eEdition 🎵 Audiobook

Strategy 1 • Use Visuals to Predict Content

Before reading, ask students to read the lesson title and look at any visuals. Then ask them to write a sentence that predicts how the visual is related to the lesson title. Repeat the exercise after reading and ask volunteers to read their sentences.

Use with Sections 1.1–1.4

Strategy 2 • Build Words from Latin Roots

Write the word *literary* on the board and underline the root *liter* in the word. Tell students that *liter* is a Latin root that means "letters" and that several words in English are built from this Latin root. Then write the related words listed below under the word *literary*. Ask a volunteer to find and underline the root *liter* in each. Ask students to suggest how each word might be related in meaning to the word *letters*.

literal	literature
literacy	illiterate

Use with Section 1.3 *The words* obliterate *and* alliteration *can also be added if appropriate for the language level of the students.*

Strategy 3 • Ask Yes/No Questions

To reinforce vocabulary meanings after reading, ask the questions below and have students say or write *yes* or *no* in response. Then reread the questions and ask students to correct the information in any sentence that has *no* as an answer (*numbers 3, 5, 6, 7, 8*).

1. Did the European Union create the euro?
2. Is the euro a form of currency?
3. Is currency a kind of tax?
4. Is the Common Market another name for the European Economic Community?
5. Is a tariff a kind of coin?
6. Are people who sell goods called consumers?
7. Is the eurozone a group of countries that use nuclear power for energy?
8. Does privatization refer to a rule that requires a citizen to serve in the military?

Use with Sections 2.1–2.3

GIFTED & TALENTED 📰 Magazine Maker

Strategy 1 • Research a Celebration

Explain to students that the celebration of the summer solstice, also known as midsummer, is an important cultural holiday in many European countries. Tell them to choose two countries that celebrate the summer solstice (such as Sweden, Finland, Iceland, Norway, or Denmark) and research information about the history and traditions of the celebrations.

Use with Section 1.1 *Invite volunteers to share their findings and, if possible, photos of the celebrations.*

Strategy 2 • Explore World Heritage Sites

Have students choose one European country from the list below and research and report on five UNESCO World Heritage Sites in that country. Tell them to describe each site and explain why it deserves the designation "World Heritage Site." Students may want to use the **Magazine Maker** for their reports.

Austria	Finland	Greece	Poland
Belgium	France	Hungary	Portugal
Czech Republic	Germany	Italy	Spain

Use with Sections 1.1–1.4

PRE-AP 🖱 Connect to NG

Strategy 1 • Develop Criteria for Comparisons

Have students use the **Research Links** to choose two capital cities in Europe and develop a set of criteria to compare them. Have students use the criteria to research each city and display their findings in a chart. Then ask them to write a paragraph saying which city they would prefer to live in and why.

Use with All Sections *Criteria might include size, population, physical features, languages spoken, and religions practiced.*

Strategy 2 • Consider Two Sides of an Issue

Have pairs of students discuss and make a chart listing the advantages and disadvantages of the entire world using the same currency. Have each pair write a paragraph stating whether they think a common currency is or is not a good idea and why.

Use with Sections 2.1 and 2.2 *You might also encourage students to use this strategy in Section 2.4 to discuss the advantages and disadvantages of setting an immigration policy.*

CHAPTER 4

EUROPE TODAY

TECHTREK FOR THIS CHAPTER

Student eEdition Maps and Graphs Interactive Whiteboard GeoActivities Digital Library Connect to NG

Go to **myNGconnect.com** for more on Europe.

PREVIEW THE CHAPTER

Buildings with traditional red-tiled roofs line this square in Prague in the Czech Republic.

TECHTREK
myNGconnect.com

Fast Forward!
Core Content Presentations
Introduce *Europe Today*

Digital Library
- GeoVideo: *Introduce Europe*
- NG Photo Gallery

Also Check Out
- Research Links in **Connect to NG**
- Graphic Organizers in **Teacher Resources**

INTRODUCE THE CHAPTER

INTRODUCE THE PHOTOGRAPH

Explain to students that Prague, the capital of the Czech Republic, is more than 1,000 years old. This photograph shows a typical view from the central and oldest part of the city. **ASK:** How would you describe the buildings that surround the square? *(Possible responses: They are narrow and three or four stories tall. There is no space between most of the buildings. They all face the square.)* Why are there tables beneath the awnings that extend from the buildings? *(They are probably tables in sidewalk cafes.)* What do you think Czechs use the square for? *(Possible response: There are no cars, but a lot of people are walking in the square. It might be a place to socialize or an open space between apartments, markets, and other businesses.)*

SHARE BACKGROUND

The city of Prague is known not only for its quaint, picturesque buildings but also for many historic structures. These include a ninth-century castle upon a hilltop, the famous Charles Bridge over the Vltava River, and several churches and cathedrals of architectural distinction. In addition, Prague is home to Charles University, which is the oldest university in Central Europe.

Encourage students to use the **Research Links** to discover more about the region.

CONNECT

The photo shows a bird's-eye view of Prague's Old Town Square, which was founded in the 10th century. One of the square's main attractions is an astronomical clock that marks every hour with the procession of 12 wooden figures.

SECTION 1 • CULTURE
How is the diversity of Europe reflected in its cultural achievements?

Four Corner Activity: Culture and Diversity This activity helps students explore the way that cultural achievements can reflect diversity by first considering the culture of the United States. Post the four items listed below in the four corners of the classroom, as indicated in the graphic. Tell the class that their task is to think about the way that U.S. culture reflects its immigrant heritage. Allow students to choose the category that most interests them and go to the appropriate corner. Ask the groups that form to brainstorm a list of culturally diverse examples related to their categories. Then have the groups share their examples with the class. `0:20` minutes

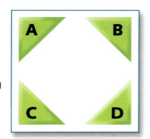

A Food and Cooking

B Music and Dance

C Stories and Literature

D Holidays and Celebrations

ACTIVE OPTIONS

NG Photo Gallery

Analyze Visuals Help students make the transition to thinking about diversity in European culture by projecting the photos of the May Day celebration, the Salzburg Music festival, and Carnival in Venice. Have students meet with a partner to discuss the following questions:

1. What types of activities seem to be part of European cultural events?

2. What cultural differences do you notice in the photos of these three events?

3. What, if anything, do the three events have in common?

After students have discussed the questions, allow them to share their ideas in class. Then ask for volunteers to predict whether the cultural achievements of Europe are more likely to reflect diversity or unity. `0:20` minutes

SECTION 2 • GOVERNMENT & ECONOMICS
What are the costs and benefits of European unification?

Jigsaw Activity: Ending Conflicts Divide the class into four expert groups and assign each group one of the following conflicts:

French Revolution and Napoleonic Wars

World War I

World War II

Cold War

Tell each group to review Section 3 of the previous chapter to answer the following questions:

1. In what way did your assigned conflict cause divisions in Europe?

2. What were some of the general causes of the conflict?

3. What events or circumstances ended the conflict?

4. What were the effects of the conflict?

After the expert groups have finished their discussion, rearrange groups using a Jigsaw strategy. See "Best Practices for Active Teaching" for a review of this cooperative learning strategy. In the new groups, each student should share the information from his or her expert group. Then the new groups should come up with a generalization about how greater unity between countries might benefit Europe and make one recommendation of a way to achieve that unity. Finally, initiate a class discussion about European unity. Ask students to consider both the advantages and disadvantages of a united Europe. `0:25` minutes

INTRODUCE CHAPTER VOCABULARY 📖 Teacher Resources

Knowledge Rating Have students complete a Knowledge-Rating chart for Key Vocabulary words. Have students copy the chart or download it for them from the **Graphic Organizers.** See "Best Practices for Active Teaching" for a review of the activity. Have students list words and fill out the chart. Then have pairs share the definitions they know. Work together as a class to complete the chart.

KEY VOCAB	KNOW IT	NOT SURE	DON'T KNOW	DEFINITION
dialect				
cuisine				
sovereignty				
assimilate				

1.1 Languages and Cultures

Main Idea Europe has a great variety of languages, cultures, and cities.

Europe has more than a half billion people, yet they live in an area that is one-half the size of the United States. In addition, the continent of Europe contains more than 40 countries. The result is a great diversity, or wide variety, of languages and cultures.

European Languages

Many of the languages spoken in Europe today fall into three language groups: Romance, Germanic, and Slavic. The Romance languages include French, Spanish, and Italian. The Germanic languages are spoken mostly in northern Europe and include German, Dutch, and English. Most people in Eastern Europe speak Slavic languages, such as Russian, Polish, and Bulgarian.

Some countries in Europe have more than one official language. Belgium, for example, has three: Dutch, French, and German. Even in countries with only one official language, people may speak different dialects. A **dialect** is a regional variety of a language. In Italy, for instance, people in Rome speak a dialect of Italian that differs from that in other cities.

Cultural Traditions

Because Europe is composed of many different countries and ethnic groups, it has a rich cultural **heritage**, or tradition. Europe's cultural diversity is reflected in its religions and celebrations.

Christianity is the dominant religion in Europe. Today, about 45 percent of the continent's total population is Catholic. Protestantism is most common in Northern Europe.

In recent years, Islam has become the fastest-growing religion in Europe. Immigrants from Turkey, North Africa, and Southwest Asia move to Europe and bring their Muslim faith with them.

Many of the holidays celebrated in Europe are rooted in religion. However, Europeans enjoy other kinds of festivals as well. One of the most colorful is the Palio, a horse race held each summer in Siena, Italy. In this race, which dates back to the Middle Ages, ten riders from ten of the city's neighborhoods compete.

City Life

More than 70 percent of Europeans live in urban areas. In Belgium, over 95 percent of the people live in or near its cities. Most of Europe's cities are **cosmopolitan**, which means that they bring together many

different cultures and influences. London is an example of a cosmopolitan city. Its restaurants and shops reflect the South Asian, Caribbean, and East Asian origins of some of its newer citizens.

Many European cities date back hundreds of years. As a result, they developed in ways very different from American cities. These cities are often smaller in area than those in the United States and have narrow, winding streets. Most people live in apartments rather than individual houses. For recreation, city dwellers visit their many parks. They also tend to use public transportation more often than most Americans.

Before You Move On
Make Inferences What might be some of the advantages and disadvantages of Europe's great variety of languages and cultures? Advantages would include a greater variety of artistic expression and a strong sense of national identities based on language and culture. Disadvantages could include difficulties communicating and an increased risk of conflict.

> **Critical Viewing** In this photo of the Palio, a centuries-old cultural tradition is honored as these horses race in Italy. What details in the photo convey the excitement of the race?
> The posture of the horses and riders, the packed stands, the gestures of the crowd, and the flying dirt all convey excitement.

ONGOING ASSESSMENT
PHOTO LAB GeoJournal

1. **Analyze Visuals** Note the way the riders are dressed in the photo. What do you suppose their clothes represent?
2. **Draw Conclusions** Study the photo and recall what you have read about the Palio. How is this race an example of a cultural tradition?
3. **Turn and Talk** Get together with a partner and discuss the holidays and festivals that you celebrate. Take notes on your discussion and be prepared to share with the rest of the class.

PLAN

OBJECTIVE
Draw conclusions about how countries preserve their languages and traditional cultures.

CRITICAL THINKING SKILLS FOR SECTION 1.1

- Main Idea
- Make Inferences
- Analyze Visuals
- Draw Conclusions
- Evaluate

PRINT RESOURCES

Teacher's Edition Resource Bank

- Reading and Note-Taking: Synthesize Ideas and Details
- Vocabulary Practice: Word Map
- **GeoActivity** Compare Urban Development

Fast Forward!
Core Content Presentations
Teach *Languages and Cultures*

Digital Library
NG Photo Gallery, Section 1

Connect to NG
Research Links

Maps and Graphs
Interactive Map Tool
Analyze Language Diversity of Europe

Also Check Out
GeoJournal in **Student eEdition**

BACKGROUND FOR THE TEACHER

Students may be surprised to learn that more Europeans speak Slavic languages than Romance or Germanic languages. Some 315 million Europeans speak Slavic languages, which include Belarusian, Bulgarian, Czech, Macedonian, Polish, Russian, Serbo-Croatian, Slovak, Slovene, and Ukrainian. Slavic languages originated in east-central Europe and spread throughout most of Eastern Europe—except Hungary and Albania. Both Albanian and Hungarian are unrelated to the three major language families.

ESSENTIAL QUESTION

How is the diversity of Europe reflected in its cultural achievements?

Section 1.1 discusses how the variety of Europe's languages, religions, and traditions demonstrates its diversity.

INTRODUCE & ENGAGE

Take a Language Survey Give students a minute to jot down the countries their families originally came from and the languages their ancestors spoke, if they know that information. If they don't know that information, invite them to write any language they can think of. Then create a class list of all the languages and circle those that are European. Ask students if they think they have listed the majority of European languages. `0:10` minutes

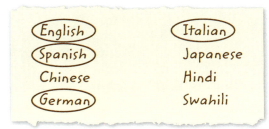

English
Spanish
Chinese
German

Italian
Japanese
Hindi
Swahili

TEACH Digital Library

Guided Discussion

1. **Evaluate** How does the existence of dialects create advantages and disadvantages for regions within larger countries? *(Dialects help give a sense of regional identity, but they also keep the region from being unified with the rest of the country.)*

2. **Make Inferences** Have students think about the festivals and celebrations in their community. Then **ASK:** What role do festivals and other celebrations play in a community? *(Festivals and celebrations help hold a community together by maintaining traditions and creating common memories.)*

Analyze Visuals Project the photos of cultural events from the **NG Photo Gallery**. Have students write a sentence or two describing the event they would most like to attend. Then ask volunteers to compare the event with one they celebrate. `0:15` minutes

DIFFERENTIATE Connect to NG

Striving Readers **Summarize** Give students the summary paragraph shown below. Have them work in pairs to fill in the blanks with Key Vocabulary. Tell them that the words are not used in the same order as in the lesson.

> More than 70 percent of Europeans live in cities. Most cities in Europe bring together many different cultures. Having many influences makes the cities _(cosmopolitan)_ . Europe has more than 40 separate countries. Dozens of ethnic groups live in Europe. These groups give Europe a rich cultural _(heritage)_ , or tradition. Many people speak a _(dialect)_ of their national language.

Gifted & Talented **Explore Culture** Have students use the **Research Links** to learn about a European cultural event such as Bastille Day in France, the Royal Edinburgh Military Tattoo in Scotland, Midsummer's Eve in Scandinavia, the Salzburg Festival in Austria, or Kiev Day in Ukraine. Then have them make a tourist poster to educate foreigners about the event and entice visitors to attend. Invite students to display their posters in class.

ACTIVE OPTIONS

Interactive Map Tool

Analyze Language Diversity of Europe

PURPOSE Compare how much language diversity European countries have

SET-UP

1. Open the **Interactive Map Tool**, set the "Region" to Europe, and set the "Map Mode" to National Geographic.

2. Under "Human Systems—Populations & Culture," turn on the Language Diversity layer. Set the transparency level to 40 percent.

3. Close the toolbar to the left and adjust the map so that Europe fills the screen.

ACTIVITY

Ask volunteers to explain the map legend. **ASK:** Are there any countries in Europe in which only one language is spoken? *(no)* Which countries have the greatest diversity of languages? *(Belgium, Bosnia and Herzegovina, Serbia)* What challenges does a country face when its citizens speak many different languages? *(Possible responses: It is harder to develop a sense of national unity. It may be difficult to find a common language for business or government transactions.)* `0:20` minutes

On Your Feet

Numbered Heads Have students get into groups of four and number off within each group. Tell the groups to discuss the ways that life in Europe is similar to and different from life in their community. After ten minutes, call a number and have a student from each group with that number share a summary of their discussion with the class. `0:20` minutes

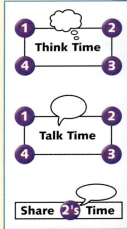

Think Time
Talk Time
Share 2's Time

ONGOING ASSESSMENT

PHOTO LAB GeoJournal

ANSWERS

1. The clothes might represent the neighborhood that each rider is representing.
2. The race represents a tradition that has continued for hundreds of years.
3. Students should discuss the customs and foods they enjoy during the holidays and festivals they observe.

1.2 Art and Music

TECHTREK
myNGconnect.com For photos of European art and samples of European music
Digital Library

Main Idea European art and music have developed over thousands of years.

Throughout the centuries, European art and music have changed to reflect different styles and beliefs.

European Art

European art grew out of the artistic achievements of ancient Greece and Rome. Greek and Roman gods and goddesses were frequent subjects of the artists from

these cultures, but they were portrayed to represent realistic human forms.

Much of the art of the Middle Ages reflected the influence of Christianity. The religious subjects were often presented as two-dimensional figures. During the Renaissance, artists used **perspective** to give their work greater depth. Although religious subjects were common, artists also painted portraits of people.

In the **Romantic period** of the early 1800s, artists moved away from religious themes to paint landscapes and other natural scenes that would convey emotion. **Impressionism** emerged in the late 1800s. Impressionist artists, such as Claude Monet, used light and color to capture a moment. By 1900, artists wanted to create a new form of art. These modern artists often worked in an **abstract** style, which emphasized form and color over realism.

Critical Viewing The *Mona Lisa*, by Italian Renaissance artist Leonardo da Vinci, is probably the most famous portrait of all time. What about this painting might account for its popularity?

Students may say that the intriguing subject, the use of perspective, and the warm colors account for the painting's popularity.

Critical Viewing This painting, *Impression, Sunrise*, by French artist Claude Monet, gave the impressionist movement its name. What kind of mood is conveyed in this painting?

Possible response: The painting conveys a calm, slightly mysterious mood.

Most opera houses contain a stage, an orchestra pit, and several levels of balconies. The ceiling of the Paris Opéra, shown here, was painted by Russian-born artist Marc Chagall.

European Music

Like art, European music began in ancient Greece and Rome. Musicians played on a few simple instruments and were often accompanied by singers.

During the Middle Ages, music was used in religious ceremonies. Singers called **troubadours** performed songs about knights and love. These songs influenced Renaissance music, when instruments such as the violin were introduced.

The new instruments helped inspire the complex rhythms in the music of the **Baroque period**, which lasted from about 1600 to 1750. **Opera**, which tells a story through words and music, was born then.

The **Classical** and Romantic periods followed the Baroque period and

continued until about 1910. Composers from these two periods, such as Ludwig van Beethoven (BAY toh vuhn) of Germany, wrote works using instruments and techniques that are still used today.

Before You Move On
Monitor Comprehension What styles and beliefs have influenced European art and music?

Religious beliefs, humanism, and changing ideas about representing nature, love, and people have influenced European art and music.

ONGOING ASSESSMENT
LISTENING LAB GeoJournal

1. **Analyze Audios** Listen to the music clip of Beethoven's Fifth Symphony in the **Digital Library**. Describe the music's mood. What instruments help to convey this mood?

2. **Form and Support Opinions** What do you think of the opening? Support your opinion by referring to specific details in the music.

PLAN

OBJECTIVE Identify the styles of music and art associated with specific periods of European history.

CRITICAL THINKING SKILLS FOR SECTION 1.2

- Main Idea
- Monitor Comprehension
- Analyze
- Form and Support Opinions
- Compare and Contrast
- Summarize

PRINT RESOURCES

Teacher's Edition Resource Bank

- Reading and Note-Taking: Sequence Events
- Vocabulary Practice: Definition Clues
- **GeoActivity** Recognize Architectural Movements

TECHTREK myNGconnect.com

Fast Forward!
Core Content Presentations
Teach *Art and Music*

Digital Library
- GeoVideo: *Introduce Europe*
- NG Photo Gallery, Section 1
- Music Clip, Section 1

Connect to NG
Research Links

Interactive Whiteboard
GeoActivity Recognize Architectural Movements

Also Check Out
GeoJournal in **Student eEdition**

BACKGROUND FOR THE TEACHER

European art became even more diverse as the period of imperialism and world trade exposed Europe to other cultures. Perhaps the most famous example is Paul Gauguin, who moved to Tahiti in 1891 to paint what he viewed as an idyllic culture. Decades earlier, Japanese prints had inspired French impressionists such as Claude Monet and Edgar Degas to use lighter colors and simpler compositions. In the 20th century, African art influenced Pablo Picasso's experiments with cubism.

ESSENTIAL QUESTION

How is the diversity of Europe reflected in its cultural achievements?

Artists and musicians from many countries have contributed to Europe's changing artistic and musical legacy. Section 1.2 describes the ways in which European art and music have evolved over the centuries.

INTRODUCE & ENGAGE 📁 Digital Library

GeoVideo: *Introduce Europe* Show students portions of the video that reflect some of the early art and architecture of Europe. As you show the video, have students raise their hands if they have seen pictures of any of these works of art or buildings before. Then invite volunteers to share what they might already know about these and other European works of art and buildings. `0:15` minutes

TEACH 📁 Digital Library

Guided Discussion

1. **Compare and Contrast** Discuss the type of music sung by troubadours in the Middle Ages. **ASK:** How were troubadours similar to and different from the singers you listen to today? *(Possible response: They both sing about love. They are different because troubadours did not have electronic instruments.)*

2. **Summarize** In what ways did European music grow more complicated over time? *(It went from simple melodies to pieces with complex rhythms. The instrumentals grew more complicated.)* Play the **Music Clip** in class. Pose the questions in the Listening Lab and discuss the answers. Invite students to share their opinions of the **Music Clip**.

MORE INFORMATION

The Piano In the early 1700s, a harpsichord designer named Bartolomeo Cristofori invented an instrument in which the strings were not plucked as on a harpsichord but struck by a hammer that instantly pulled back when the key was released. This mechanism allowed players to vary the length and volume of notes. Because it was a more expressive instrument than its predecessor, the piano became favored by European composers.

Piano, c. 1720

DIFFERENTIATE 📱 Connect to NG

English Language Learners **Match Words and Definitions** Give students the following matching exercise. Have them work in pairs to match the words with their definitions and then write a sentence using each one.

(d)	**1.** Romantic period	a.	sang about knights and love
(e)	**2.** impressionism	b.	uses color and form, not realism
(b)	**3.** abstract	c.	tells a story through words and music
(a)	**4.** troubadour	d.	uses scenes of nature to convey emotion
(c)	**5.** opera	e.	uses color and light to capture a moment

Pre-AP **Write Reports** Have students use the **Research Links** to learn about important people and works in the field of European art or music. Have them choose one person and write a review of the artist's or musician's work. Students should imagine that they are music or art critics and support their critiques citing details from their subject's works and facts about his or her life.

ACTIVE OPTIONS

📁 NG Photo Gallery

Analyze Visuals Have students view the photos of European art. Invite students to share their opinions of the artwork. **ASK:** What, if any, elements in these early works of art do you recognize in any later works you might have seen? *(recognizable forms, use of color, use of perspective, creation of a particular mood)* `0:15` minutes

On Your Feet

Build a Paragraph Have students form five lines. Give each group the same topic sentence: *The traditions of European music shape the music we hear in movies and on television today.* Tell them to build a paragraph by having each student in the line add one sentence giving an example of how European musical traditions affect soundtracks. Groups can present their paragraphs to the class with each student reading his or her sentence. `0:15` minutes

📋 Interactive Whiteboard
GeoActivity

Recognize Architectural Movements Have students read the passage individually and then work in small groups to identify the architectural movement each building represents. Allow groups to compare their answers with other groups and make corrections where necessary. `0:20` minutes

EXTENSION Challenge students to think about the architectural styles represented in buildings in their community. Have them discuss whether they see any of the European architectural movements represented. `0:10` minutes

ONGOING ASSESSMENT
LISTENING LAB 📋 GeoJournal

ANSWERS

1. Possible responses: The mood is dark and powerful. Beethoven uses violins, cellos, oboes, and horns to achieve this mood.
2. Possible responses: I liked it because the opening notes were repeated in a way that was very dramatic. I didn't like it because there were so many instruments playing that it was complicated and hard to listen to.

1.3 Europe's Literary Heritage

TECHTREK
myNGconnect.com For photos of
European writers and Guided Writing
Digital Library • Student Resources

Main Idea European literature has reflected new ways of thinking over the centuries.

Plays by the English playwright William Shakespeare (1564–1616) are performed almost every day. European writers such as Shakespeare have influenced literature for centuries. They wrote in many different genres, or forms of literature, including poetry, plays, and novels.

Literary Origins

European literature began with the ancient Greeks and Romans. Around 800 B.C., the Greek poet Homer wrote the epic poems *The Iliad* and *The Odyssey*. An epic poem is a long poem that tells the adventures of a hero who is important to a particular nation or culture. Around 20 B.C., the Roman poet Virgil wrote *The Aeneid*, an epic poem about the founding of Rome.

One of the greatest writers of the late Middle Ages and early Renaissance was the Italian poet Dante (1265–1321). As you have learned, Dante wrote *The Divine Comedy* in Italian, not in Latin. The epic poem deals with the religious beliefs and politics of his time.

Many later works of the Renaissance focused on human behavior. Shakespeare explored this theme in plays such as *Hamlet*. Spanish writer Miguel de Cervantes (1547–1616) wrote what is considered the first modern novel, *Don Quixote* (kee HO tee). A novel is a long work of fiction, containing characters and a plot. The printing press, which Johannes Gutenberg developed in the 1450s, helped spread the popularity of these books.

The 1700s and 1800s

In the mid-1700s, Enlightenment ideas about reason and government inspired the movement toward democracy. These ideas, in turn, led French and English writers of the time, such as Voltaire and John Locke, to explore the rights of the individual.

In the 1800s, writers of the Romantic period continued this exploration, with an emphasis on emotion and nature. For example, German author Johann Wolfgang von Goethe (GHER tuh) (1749–1832) wrote *The Sorrows of Young Werther*, a novel about a sensitive young artist.

Other writers of the 1800s took a much more realistic look at life. In novels such as *Sense and Sensibility*, British writer Jane Austen (1775–1817) used humor to examine women's role in society.

Critical Viewing Inspired by tales of knights, Don Quixote (right) goes to battle evil with his servant Sancho Panza (left). What details in this painting suggest that Cervantes' novel is a comedy? The contrast between tall, thin Don Quixote dressed in armor and his humble, round-bellied servant is humorous, as is the fact that Sancho Panza rides a donkey.

Critical Viewing Austen's novels typically end with marriage, such as the one shown in this scene from a film adaptation of *Sense and Sensibility*. Based on the photo, how would you describe an English wedding in the 1800s? Possible response: modest, joyous, festive, religious

Another British writer, Charles Dickens (1812–1870), commented on social issues, including poverty, in such novels as *Oliver Twist*. Norwegian playwright Henrik Ibsen (1828–1906) wrote plays, such as *A Doll's House*, which criticized the traditional role of husbands and wives at that time.

Modern Literature

The two world wars had a great impact on the modern literature of the 20th century. Writers at this time reflected the sense that life was uncertain and unpredictable. Some rejected traditional genres and experimented with writing new forms of plays, poems, and novels.

Many modern writers examined the inner workings of the mind. In the novel *Ulysses*, for example, Irish writer James Joyce (1882–1941) focused on the thought processes of the main character over the course of a single day. Romanian playwright Eugene Ionesco (1909–1994) used ridiculous situations to comment on

what he saw as the emptiness of life. Many European and other writers today have been influenced by these authors.

Before You Move On
Summarize What new ways of thinking has European literature reflected over the centuries?

European literature has reflected changing ideas about religion, the role of the individual, society, women, and attitudes toward life.

ONGOING ASSESSMENT
WRITING LAB
GeoJournal

Write Reports Think about what you have learned about Europe's literary heritage. Then consider the following question: In what ways do beliefs and events influence literature? Select a writer mentioned in this lesson and write a report in which you answer this question.

Step 1 Research to learn more about the writer and the period in which he or she lived.

Step 2 Find out how the beliefs and events of the time influenced the writer.

Step 3 Write a brief report in which you explain these influences. Support your ideas with specific references to one or two of the writer's works. Go to **Student Resources** for Guided Writing support.

PLAN

OBJECTIVE Explain the relationships between literary movements and historical events.

CRITICAL THINKING SKILLS FOR SECTION 1.3

- Main Idea
- Summarize
- Write Reports
- Compare and Contrast
- Analyze Effects
- Form and Support Opinions
- Create Time Lines

PRINT RESOURCES

Teacher's Edition Resource Bank

- Reading and Note-Taking: Find Main Idea and Details
- Vocabulary Practice: Compare/Contrast Paragraph
- **GeoActivity** Analyze Primary Sources: Romantic Writing

TECHTREK myNGconnect.com

» **Fast Forward!**
Core Content Presentations
Teach *Europe's Literary Heritage*

Interactive Whiteboard
GeoActivity Analyze Primary Sources: Romantic Writing

Also Check Out
- NG Photo Gallery in **Digital Library**
- Writing Templates in **Teacher Resources**
- GeoJournal in **Student eEdition**

BACKGROUND FOR THE TEACHER

Many European classics are accessible to middle school students in terms of subject matter and provide insight into the times in which they were written. The following works are in the public domain and can easily be found online: *Peter Pan* by J.M. Barrie; *The Secret Garden* by Frances Hodgson Burnett; *Alice's Adventures in Wonderland* by Lewis Carroll; *Robinson Crusoe* by Daniel Defoe; *Kim* by Rudyard Kipling; *Black Beauty* by Anna Sewell; *The Swiss Family Robinson* by Johann David Wyss.

ESSENTIAL QUESTION

How is the diversity of Europe reflected in its cultural achievements?

Section 1.3 reflects on the fact that European literary traditions include a diversity of genres, movements, individual voices, and perspectives on national traditions as well as on what it means to be human.

INTRODUCE & ENGAGE

Activate Prior Knowledge Have students name novels they have read that were set in a country other than the United States or set in a historical time period. **ASK:** What have you learned about other cultures from reading novels? Discuss students' ideas and create a list of books that students feel would be valuable in helping readers experience other cultures. **0:10** minutes

TEACH

Guided Discussion

1. **Compare and Contrast** How are the poems *The Aeneid* and *The Divine Comedy* similar and different? *(Both are epic poems. Their subject matter is different:* The Aeneid *is about the founding of Rome while* The Divine Comedy *is about religious beliefs.)*

2. **Analyze Effects** How did the two world wars of the 20th century affect literature? *(Writers began to portray life as unpredictable and uncertain.)*

3. **Form and Support Opinions** Initiate a discussion about students' favorite novels or stories and what they have learned or liked about them. **ASK:** What is the purpose of literature? Support your opinion with details from the text. *(Possible response: Literature has many purposes, but one important purpose is to convey ideas and to help people understand community issues. For example,* The Divine Comedy *helps readers understand religious beliefs and the politics of the time in which it is set.)*

Create Time Lines Have pairs of students work together to create a time line of European literature using the periods, writers, and titles discussed in the lesson. Tell students that their time lines should begin with 800 B.C. and end with the 1900s. Encourage students to refer to the time line as they study the material in the lesson. **0:20** minutes

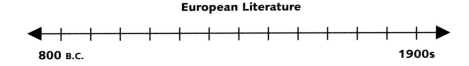

European Literature

800 B.C.　　　　　　　　　　　　　　　　　1900s

DIFFERENTIATE

Inclusion Compare Genres Students who are visual learners may benefit from seeing examples of the genres mentioned in the lesson: novels, poetry, and plays. Show sample pages of each genre so that students can compare the different formats. Invite students to discuss the similarities and differences among the genres.

Pre-AP Create Literary Maps Give students an outline map of Europe. Suggest that students choose four or five countries and research to learn about a major author from each country. Create callouts for the map with each author's picture, list of important works, and brief biographical information, such as dates of birth and one or two important facts.

ACTIVE OPTIONS

Interactive Whiteboard
GeoActivity

Analyze Primary Sources: Romantic Writing Have students work in pairs. Partners should take turns reading the excerpts. Then they should work together to answer the critical thinking questions. Allow students to compare their answers with those of another pair and make corrections where necessary. **0:20** minutes

EXTENSION Have pairs brainstorm a list of contemporary music, novels, and films that they think are influenced by the Romantic period. Allow students to share their best examples in class. **0:15** minutes

On Your Feet

Literature Roundtable Divide the class into groups of four or five. Have the groups move desks together to form a table where they can all sit. Hand each group a sheet of paper with the question *How does European literature reflect the diversity of European culture?* The first student in each group should write an answer, read it aloud, and pass the paper clockwise to the next student. The paper should circulate around the table until students run out of answers or time is up. **0:25** minutes

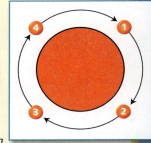

WRITING LAB GeoJournal

ANSWERS

Students' reports should show the following qualities:

1. The introduction should clearly state the topic—a writer mentioned in the section—and identify the culture and time period in which the writer lived.

2. The body of the report should describe how beliefs and events of the time period influenced the writer.
   a. First, the report should identify which beliefs or events shaped the writer's work.
   b. Second, the report should describe specific examples from one or two works that demonstrate this influence.

3. The conclusion should offer a generalization about how culture influences literature and then briefly summarize the evidence offered in the body in support of that generalization.

1.4 Cuisines of Europe

TECHTREK

myNGconnect.com For photos
of traditional European foods

Digital
Library

Main Idea Landforms and climate have influenced the cooking traditions of Europe.

Throughout most of Europe, foods such as meat, bread, and cheese are ==staples==, or basic parts of people's diets. However, the ==cuisine==, or cooking traditions, of most European countries is largely determined by the landforms and climate of a particular region.

Foods of Western Europe

The hot, dry climate in the Mediterranean countries of Spain, France, Italy, and Greece is ideal for growing olives, tomatoes, and garlic. As a result, these are key ingredients in their cuisines. Fish from the bodies of water surrounding the countries is also an important menu item.

The cuisines of France and Italy have influenced cooking throughout the world and are especially known for their sauces. French sauces are typically made of milk or cheese, while those of Italy are often tomato based.

People in western European countries with cooler climates often eat more filling fare. In Germany, Great Britain, and Ireland, potatoes grow well and are popular side dishes. In contrast with the light sauces of France and Italy, German cooks often serve heavier gravies.

In Scandinavian countries such as Sweden and Norway, the people often eat herring and other fish. Scandinavians also enjoy deer meat provided by the herding culture of these northern countries.

Foods of Eastern Europe

Eastern Europe's cold climate results in a shorter growing season than that of Western Europe. In Russia, root vegetables such as turnips and beets are well adapted to the country's climate. A soup called *borscht*, which is made from beets, is a traditional offering on cold winter nights.

The fertile soil of Hungary allows Hungarian farmers to grow grains and potatoes. These crops are used to make a variety of breads and dumplings. A meat stew called *goulash* is Hungary's national dish. It is made with beef, potatoes, and vegetables, and seasoned with paprika. Paprika is a red spice that was brought to Hungary by the Turks in the 1500s.

Bread like this round loaf is the traditional centerpiece at a Ukrainian wedding.

Like Hungary, Ukraine has fertile soil and fields of wheat and other grains. The country is known for its bread. On special occasions, cooks prepare breads decorated with ornaments made of dough.

Before You Move On
Summarize How have landforms and climate influenced European cooking traditions? Climate, landforms, and location have determined what countries grow, harvest, and eat.

> **Critical Viewing** The French often enjoy long, relaxed meals with friends and family. What attitude toward life and food does this traditional way of eating suggest?
a love of food, a relaxed attitude toward life, an emphasis on relationships

ONGOING ASSESSMENT
SPEAKING LAB GeoJournal

Turn and Talk What is the traditional cuisine of the United States? What impact have dishes brought by immigrants from other countries had on how and what Americans eat? Get together in a small group and discuss these questions. Be prepared to share your ideas with other groups.

PLAN

OBJECTIVE Summarize the characteristics of the different regional cuisines of Europe.

CRITICAL THINKING SKILLS FOR SECTION 1.4

- Main Idea
- Summarize
- Find Main Idea
- Synthesize
- Analyze Visuals

PRINT RESOURCES

Teacher's Edition Resource Bank

- Reading and Note-Taking: Categorize Cuisines
- Vocabulary Practice: Comparison Chart
- **GeoActivity** Graph Olive Oil Production Rates

TECHTREK myNGconnect.com

Fast Forward!
Core Content Presentations
Teach *Cuisines of Europe*

Interactive Whiteboard
GeoActivity Graph Olive Oil Production Rates

Digital Library
NG Photo Gallery, Section 1

Also Check Out
- Graphic Organizers in **Teacher Resources**
- GeoJournal in **Student eEdition**

BACKGROUND FOR THE TEACHER

Many European cities and countries are known for making special varieties of sausage using recipes that date back to the Middle Ages. To make sausage, cooks stuff a mixture of finely chopped meat, seasonings, and other ingredients into casings. Some traditional European sausages, such as bologna and frankfurters, have been adopted as part of American cuisine. Others, such as eastern European blood sausage and Swedish potato sausage, are less widely eaten in the United States.

ESSENTIAL QUESTION

How is the diversity of Europe reflected in its cultural achievements?

The cuisines of Europe reflect the products and traditions of its different regions. Section 1.4 describes the ways in which European foods are connected to geography and culture.

INTRODUCE & ENGAGE

Create Agricultural Maps Ask students if they or their relatives have ever grown a vegetable garden. Have students with knowledge of gardening explain how their local climate affects what plants can be grown. Then have students use the climate map in Section 1.1 of the previous chapter to help them create a map showing where vegetables would grow best in Europe. Have students use an outline map of Europe to draw their maps. Once students have completed the maps, **ASK:** In what parts of Europe are people more likely to grow vegetables? How might that affect their cooking traditions? *(in the areas with a humid temperate climate; countries with this climate may be more likely to emphasize the use of fresh vegetables in their cuisine)*
`0:20` minutes

TEACH   Digital Library

Guided Discussion

1. **Find Main Idea** How has the geography of Hungary and Ukraine affected the cuisine of those two countries? *(Both countries have fertile soils and grow a great deal of grain, so dumplings and bread are important foods there.)*

2. **Synthesize** While the cuisine of each country reflects its geography and culture, do the people who live in Europe now always consume only their regional specialties? Why or why not? *(Possible response: No, while regional fare is important, modern transportation and communication have allowed cuisines to be shared around the world.)*

Analyze Visuals Project the photos of European foods from the **NG Photo Gallery**. For each photograph of a prepared dish, **ASK:**

1. What type of food is shown here? What are some of the ingredients?
2. How much skill do you think a person needs to prepare this dish?
3. Does this remind you of any of the foods you eat? `0:15` minutes

DIFFERENTIATE

English Language Learners Understand **Multiple-Meaning Words** Point out the Key Vocabulary word *staple*. Students are probably more familiar with the use of that word to mean a small folded wire used to fasten sheets of paper together. Ask students to draw images to remind them of the two meanings of this word.

Have an English learner work with a partner who is fluent in English to review the section and find at least two other multiple-meaning words. Ask them to make a Word Web for each word, giving its different definitions.

Gifted & Talented **Compile a Cookbook** Have groups of students compile a cookbook that showcases foods from all over Europe. Ask them to include brief notes for each recipe on why the dish is made in a particular country or area. If possible, students should also illustrate the recipes with photos or drawings.

staple

ACTIVE OPTIONS

Interactive Whiteboard
GeoActivity

Graph Olive Oil Production Rates After students have read the passage on their own, have them work in groups to calculate the increases in olive oil production in each country and then use their calculations to create a bar graph. You might explain what it means when a percentage exceeds 100. After the graphs are completed, have the groups answer the critical thinking questions in the activity. `0:20` minutes

On Your Feet

Three-Step Interview Have students choose a partner. One student should interview the other on the question *What European cuisine would you most like to try? Why?* Tell students that they should name specific dishes that they would like to eat. Then they should reverse roles. Finally, each student should share the results of his or her interview with the class. `0:15` minutes

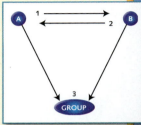

Performance Assessment

Write Journal Entries Show students the photos of European culture from the **NG Photo Gallery**. Discuss the photos, inviting students' comments and reactions. Then have them use the photos to write journal entries describing what they experienced each day as they traveled through Europe on a one-week tour, with each day being spent in a different country. Go to **myNGconnect.com** for the rubric.

ONGOING ASSESSMENT
SPEAKING LAB   GeoJournal

ANSWERS
Students might note such traditional American foods as fried chicken and acknowledge the contributions of immigrants from countries such as Mexico, China, Italy, and India.

2.1 The European Union

TECHTREK
myNGconnect.com For an online map and news about the European Union

Maps and Graphs • Connect to NG

Main Idea The European Union was formed to unite Europe and benefit it economically.

In 1948, the United States established a program called the Marshall Plan to help Europe rebuild after World War II. To manage the U.S. aid money, European countries formed the Organization for European Economic Cooperation in 1948. As a result, European countries discovered that they could rebuild their countries faster when they worked together.

The Common Market

In 1957, some European countries sought even closer economic ties. They formed the European Economic Community (EEC), which became known as the **Common Market**. The first countries to join were Belgium, France, Italy, Luxembourg, the Netherlands, and West Germany. Several more countries joined during the 1980s.

The Common Market pledged to create "an ever closer union among the European peoples." However, it was primarily formed to create a single market among the member nations. A single market is one in which a group of countries trades across its borders without restrictions or tariffs. A **tariff** is a tax paid on imports and exports.

A United Europe

In 1992, the countries of the Common Market sought to extend their economic organization throughout Europe. They met in Maastricht in the Netherlands and signed the Treaty of Maastricht, which created the **European Union (EU)**. By 2010, the EU had 27 member nations with a total of more than 500 million people. When considered as a single economy, the EU is the largest in the world.

EUROPEAN UNION

MAP TIP On this map, the year beneath a country's name indicates when it joined either the EEC (between 1957 and 1991) or the EU (beginning in 1992).

Legend:
- Member
- Candidate
- Potential candidate

The EU has a government with an executive, a legislative, and a judicial branch. These branches propose, pass, and enforce the organization's policies and legislation. The EU also has agencies that direct economic policies. Through these agencies, the EU has eliminated tariffs among most member nations and founded a European bank. In 1999, the EU created a common **currency**, or form of money, called the **euro**. By 2011, 17 member nations had adopted this currency.

One of the requirements for joining the EU is having a stable democracy that respects human rights. Some countries that have applied for membership, such as Turkey, are still under review. However, other European countries, including Norway, have chosen not to join. Norway does not want to give up its **sovereignty**, or control over its own affairs.

Before You Move On
Monitor Comprehension How has the European Union helped unite Europe?

ONGOING ASSESSMENT
MAP LAB GeoJournal

1. **Interpret Maps** Study the map and recall what you have learned about Europe's geography. In what ways does Europe's geography encourage unity?

2. **Draw Conclusions** Based on the map, in what area are most of the members who joined in later years located? What political situation might have prevented them from joining earlier?

SECTION 2.1 **125**

Critical Viewing The EU flag, seen here in front of the organization's Parliament building in Strasbourg, France, is also the flag of Europe. The flag's circle of stars represents European unity. Why is this a fitting symbol for the EU?

European Union flag

The EU was formed to create unity among European nations.

The EU has eliminated tariffs among most member nations and created a common currency. It also promotes democracy on the continent.

PLAN

OBJECTIVE Explain why European countries with different interests chose to form the European Union.

CRITICAL THINKING SKILLS FOR SECTION 2.1

- Main Idea
- Monitor Comprehension
- Interpret Maps
- Draw Conclusions
- Make Inferences

PRINT RESOURCES

Teacher's Edition Resource Bank

- Reading and Note-Taking: Outline and Take Notes
- Vocabulary Practice: Words in Context
- **GeoActivity** Categorize Fundamental Rights

TECHTREK myNGconnect.com

Fast Forward!
Core Content Presentations
Teach *The European Union*

Digital Library
GeoVideo: *Introduce Europe*

Connect to NG
Research Links

Maps and Graphs
- **Interactive Map Tool** Explore EU Membership Patterns
- Online World Atlas: European Union

Also Check Out
- NG Photo Gallery in **Digital Library**
- GeoJournal in **Student eEdition**

BACKGROUND FOR THE TEACHER

According to the European Union Web site, the EU is neither a federation nor a simple cooperative organization: "The countries that make up the EU (its 'member states') remain independent sovereign nations but they pool their sovereignty in order to gain a strength and world influence none of them could have on their own."

ESSENTIAL QUESTION

What are the costs and benefits of European unification?

Section 2.1 explains that, after World War II, western European countries began forming closer economic ties that led to greater political ties.

INTRODUCE & ENGAGE Digital Library

GeoVideo: *Introduce Europe* Show students the portion of the video that features the European Union. Ask students why the countries of Europe might want to unite economically and politically. `0:10` minutes

TEACH Maps and Graphs

Guided Discussion

1. **Draw Conclusions** Why would eliminating tariffs help trade between countries? (*A tariff is a tax on imports and exports, so it makes trade more expensive. Reducing the cost of trade makes it more profitable and so leads to increased trade.*)

2. **Make Inferences** Why do you think the European Union requires its members to have democratic governments that respect human rights? (*Possible responses: It may believe that such governments will help prevent another destructive European war. It may believe that such governments are most compatible with strong economies.*)

3. **Interpret Maps** Project the European Union map from the **Online World Atlas.** Have students study the map. **ASK:** How is Turkey geographically different from the other candidates for EU membership? (*Most of its territory lies in Asia.*) Then ask students if they think Turkey's location might be a determining factor in its possible membership.

MORE INFORMATION

EU Government Three main institutions make up the government of the European Union. The European Commission, which represents the Union as a whole, proposes new laws and executes laws that pass. Once proposed, laws must be adopted by the European Parliament, which is elected by European citizens, and the Council of the European Union, which is composed of ministers that represent member states. In addition, the EU has a Court of Justice that upholds the law and a Court of Auditors that oversees finances.

DIFFERENTIATE Connect to NG

English Language Learners **Create Vocabulary Cards** Have students create flash cards for several of the important economic words in this section: *tariff, import, export, currency, euro.* Suggest that they write each word on the front of a card. The back can have an English definition, the corresponding term from their home language, or a picture that conveys the word's meaning.

Pre-AP **Stage a Debate** Have students use the **Research Links** to find out why Norway and Switzerland have decided to stay out of the European Union. Suggest that they look for political, economic, and cultural reasons. Then have students stage a debate in which some students support the reasons why Norway and Switzerland should join the EU and other students present the argument against joining.

ACTIVE OPTIONS

Interactive Map Tool

Explore EU Membership Patterns

PURPOSE Identify which nations joined the European Union during certain decades

SET-UP

1. Open the **Interactive Map Tool**, set the "Region" to Europe, and set the "Map Mode" to Topographic.

ACTIVITY

1. Have the class review the European Union map in their text. Ask students to identify the original members of the European Union. (*Germany, France, Italy, Belgium, Luxembourg, the Netherlands*) Click on the DRAWING TOOLS tab and ask a volunteer to use the Free-form Line Tool to draw an outline around those original members. Make sure that the volunteer also outlines the border between those countries and nonmember Switzerland.

2. Have a second volunteer choose a different outline color and use the Free-form Line Tool to outline the group of countries that joined during the 1970s.

3. Repeat the process for the countries that joined in the 1980s, the 1990s, and the 2000s, using a new color each time. `0:20` minutes

On Your Feet

Thumbs Up/Thumbs Down Divide the class into groups and have each group write six true-false statements about the lesson with the correct answers included. Collect the questions. Mix them up and read them aloud to the class, skipping any duplicates. Have students give a "thumbs up" for true statements and a "thumbs down" for false statements. Correct any misconceptions. `0:25` minutes

ONGOING ASSESSMENT
MAP LAB GeoJournal

ANSWERS

1. Europe is composed of relatively small countries that share borders and lie close together.

2. Most of the members who joined later were in Eastern Europe. Many eastern European countries were satellites of the Soviet Union until the democratic revolutions of the late 1980s and the collapse of the Soviet Union in 1991. Their membership would have been delayed as they worked to establish democratic governments and market economies.

2.2 The Impact of the Euro

TECHTREK
myNGconnect.com For photos of the euro

Digital Library

Main Idea The euro has helped to unify Europe both economically and politically.

As you have learned, the European Union (EU) created the euro in 1999. Since then, many of the member nations have adopted the euro, which has had a significant impact on Europe.

The Euro Arrives

The euro was launched in 1999, but paper money and coins of the currency were not issued until 2002. The 17 countries that use the euro are known as the eurozone. Some countries that belong to the EU, including Romania, hope to join the eurozone soon. Other countries, including Great Britain and Denmark, have not adopted the euro. They believe that giving up their own currency might result in a loss of control over their economies.

The symbol for the euro is €. Euro paper money, or banknotes, is the same throughout the eurozone. Euro coins, however, differ from country to country. The front, or common, side of each coin has the same image and a number indicating its value. The back, or national, side shows a design that was chosen by the member nation.

Economic Benefits of the Euro

The euro allows people, money, and goods to move freely within the eurozone. Before the creation of the euro, when a French citizen traveled to Germany, for example, he or she had to pay a fee to exchange, or convert, francs—the French currency—into marks—the German currency. Because the common currency has made travel easier and less expensive, tourism has increased within much of Europe.

A single currency means lower fees for conducting business. As a result, trade has increased among European nations by an estimated 10 percent since 2002. The currency has also made costs easier to compare for companies within the eurozone. As a result, companies can import the least expensive products and then pass along the savings to consumers, the people who buy the goods.

Political Benefits of the Euro

In 2010, the unity of the eurozone was tested. Greece and Ireland—countries in the eurozone—were deeply in debt. To help them manage their debt, the other eurozone nations loaned the two countries money. In return, Greece and Ireland had to raise taxes and reduce spending. By cooperating, the eurozone was able to help two of its member nations.

1-EURO COIN OF THREE EUROZONE COUNTRIES

Front   Back

AUSTRIA Mozart, Austrian composer

GERMANY Eagle, German symbol

IRELAND Harp, Irish symbol

Before You Move On
Summarize In what ways has the euro helped Europe unify economically and politically? It has facilitated the movement of people, money, and goods.

ONGOING ASSESSMENT
VIEWING LAB GeoJournal

1. **Compare and Contrast** Study the euros in the diagram. Note that the front of the coin is the same for all three countries shown. What elements appear on the front of each euro? What does each element represent?

2. **Make Inferences** Why might the countries have wanted their own design on the euro?

3. **Conduct Internet Research** Go online to find out what impact helping Greece and Ireland has had on the euro and the eurozone. Share your findings with the class.

The prices of fruit in this Italian market are in euros.

SECTION 2.2 **127**

PLAN

OBJECTIVE Draw conclusions about the effects of a unified currency on the European Union.

CRITICAL THINKING SKILLS FOR SECTION 2.2

- Main Idea
- Summarize
- Compare and Contrast
- Make Inferences
- Conduct Internet Research
- Form and Support Opinions
- Analyze Visuals

PRINT RESOURCES

Teacher's Edition Resource Bank

- Reading and Note-Taking: Draw Conclusions
- Vocabulary Practice: Travel Blog
- **GeoActivity** Graph the Economic Impact of the Euro

TECHTREK myNGconnect.com

Fast Forward!
Core Content Presentations
Teach *The Impact of the Euro*

Teacher Resources
Graphic Organizers

Digital Library
NG Photo Gallery, Section 2

Interactive Whiteboard
GeoActivity Graph the Economic Impact of the Euro

Also Check Out
GeoJournal in **Student eEdition**

BACKGROUND FOR THE TEACHER

Because the combined EU economy is the world's largest, the euro has become an important international currency, second only to the U.S. dollar. In many cases, international debt is calculated in euros, and many global loans are made in euros. In international trade, some companies choose to issue their invoices and pay their bills in euros. In addition, some countries that are not EU members, such as Monaco, San Marino, and Vatican City, use the euro as their official currency.

ESSENTIAL QUESTION

What are the costs and benefits of European unification?

The euro has served to unify Europe financially and has also served as a symbol of that unity. Section 2.2 discusses the background and benefits of the currency.

INTRODUCE & ENGAGE   📖 Teacher Resources

K-W-L Chart Provide each student with a K-W-L chart downloaded from the **Graphic Organizers**. Have students recall what they have already learned about the European Union. Then have them preview Section 2.2 by looking at the visuals and their captions. Ask students to write questions about the euro that they would like to have answered as they study the section. Allow time after students have finished reading to record what they learned. `0:15` minutes

TEACH   🖥 Digital Library

Guided Discussion

1. **Summarize** Have students reread the two paragraphs under "Economic Benefits of the Euro." **ASK:** How has the euro affected individual people? *(It has made tourism easier so more people have traveled to other EU countries. It has allowed consumers to pay less for goods because companies have found it easier to find and offer the cheapest products.)*

2. **Form and Support Opinions** Do you think the countries of North America would benefit from having a single currency? Why or why not? *(yes, because a single currency would make travel and trade easier; no, because the United States would end up supporting North American countries with much weaker economies)*

Analyze Visuals Project the photos of euros from the **NG Photo Gallery**. Identify and discuss the designs on the national side of the coins. Then ask students what they can conclude about the culture of each country from the designs. `0:10` minutes

DIFFERENTIATE

Striving Readers Analyze Cause and Effect Give students a graphic organizer like the one below. Have them work in pairs to find the effects of adopting the euro and record them in the Effect boxes.

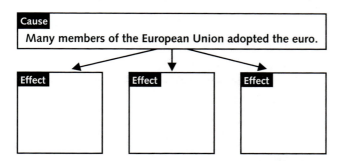

Cause
Many members of the European Union adopted the euro.

Effect   **Effect**   **Effect**

Inclusion Make Connections To help students who learn best from concrete examples, bring in several U.S. quarters featuring different states or territories or find photos that you can project. Put the coins in a line, heads up, on a desk or table. Then turn each quarter over, showing the commemorative designs on the back. **ASK:**

- Do the different designs affect the value of the quarters or their ability to be spent in any state? *(no)*
- How are these quarters like the euro coins shown in the book? *(They are a shared currency. They have special designs honoring member states.)*
- How are the euros different from quarters? *(They are issued by an international organization, not a single country.)*

Interactive Whiteboard
GeoActivity

Graph the Economic Impact of the Euro
Have students complete the graph in small groups to understand the changes to GDP per capita and exports after countries adopted the euro. Monitor groups as they plot points on the graph. Allow groups to compare their graphs with those of other groups and make corrections where necessary. As a class, discuss the graphs and then answer the critical thinking questions. `0:20` minutes

On Your Feet

Take a Stand On one wall of your classroom, post a sign that says, "All EU members should adopt the euro." On the opposite wall, post a sign that says, "All EU members should keep their own national currencies." Tell students they can adopt either of those opinions or an opinion that is some combination of the two. Tell students to find the position on a line between the two signs that best represents their opinion. Have students explain the reasons for their stance by citing information from the lesson. `0:15` minutes

On Your Feet

Shop Using Euros Write the current value of the euro (in relation to the dollar) on the board. Then form half the class into small groups and have them set up grocery stores around the classroom. Group members should write each item they are selling on a separate piece of paper and indicate its price in euros. Ask each of the remaining students to use small pieces of paper to make play money totaling 50 euros. Then have these students use their euros to shop at the grocery stores. After students have finished shopping, have them work with partners to figure out how much their purchases cost in dollars. `0:20` minutes

VIEWING LAB   📝 GeoJournal

ANSWERS

1. The front of each coin shows a map of Europe and 12 stars, which appear on the EU flag and symbolize Europe's unity.
2. They have pride in their culture, country, and heritage.
3. Responses will vary, but students may find that helping Greece and Ireland weakened the value of the euro and endangered the existence of the currency.

2.3 Democracy in Eastern Europe

TECHTREK
myNGconnect.com For photos reflecting democratic progress in Eastern Europe
Digital Library

Main Idea Eastern European countries have faced many challenges in their transition to democracy.

After World War II, many eastern European countries came under the control of the Soviet Union. The citizens of these countries lacked democratic freedoms and had a low standard of living. In 1981, Poland rebelled peacefully against its Communist government. By the late 1980s, similar rebellions had spread throughout Eastern Europe. Finally, in 1991, Russia and several other republics declared their independence, and the Soviet Union collapsed.

The Road to Democracy

Since gaining their independence, Poland, Hungary, and the Czech Republic developed stable democratic governments In other countries, **democratization**, or the process of becoming a democracy, has been more difficult to achieve. In 1991, civil war broke out among ethnic groups in Yugoslavia. Over time, the country divided into several new democratic countries, including Serbia and Croatia.

Ukraine has also had setbacks. In 2004, the Ukrainian people staged the **Orange Revolution** and peacefully removed their prime minister, Viktor Yanukovych.

Many believed that he was corrupt and was being controlled by Russia. However, their new leader, Viktor Yushchenko, disappointed the Ukrainians. Some believed he had become anti-democratic and blamed him for their weakened economy. In 2010, the voters brought Yanukovych back to power.

Rebuilding Economies

The former Communist countries of Eastern Europe also began to rebuild their economies. They changed from government-controlled economies to market economies. They accomplished this goal through **privatization**. That means that government-owned businesses became privately owned.

Eastern European countries have had mixed results since making the adjustment to a market economy. Poland has had the greatest success. It has a fast-growing economy and exports goods throughout Europe. Other countries have been slower to establish new businesses and become competitive. They have also experienced rises in prices and unemployment.

The leaders of many eastern European countries wish to integrate with the rest of Europe. They want to join the European Union and NATO, a military alliance of democratic states in Europe and North America. While some citizens of eastern European countries believe that they were more secure under Communist leaders, others—particularly young people—disagree. They favor democracy and feel that this form of government can better help them solve their countries' problems.

Before You Move On

Summarize What challenges have eastern European countries faced in their transition to democracy and a market economy? completely overhauling their government, adjusting to a market economy

ONGOING ASSESSMENT
READING LAB GeoJournal

1. **Identify Problems and Solutions** What problem did Ukraine face in its transition to democracy? In what way did the solution to the problem reflect democratization?

2. **Make Inferences** Why do you think younger eastern Europeans might be more willing than older people to support the democratic movements in their countries?

Critical Viewing Polish citizens shop and relax in a spacious mall in Warsaw. What does the shopping complex in this photo suggest about Poland's economy?
The modern shopping complex suggests that Poland's economy is strong and that the Polish people have money to spend.

SECTION 2.3 **129**

PLAN

OBJECTIVE Explain the challenges that eastern European countries face in making the transition from communism to democracy.

CRITICAL THINKING SKILLS FOR SECTION 2.3

- Main Idea
- Summarize
- Identify Problems and Solutions
- Make Inferences
- Synthesize
- Explain
- Analyze Visuals

PRINT RESOURCES

Teacher's Edition Resource Bank

- Reading and Note-Taking: Summarize Governments and Economies
- Vocabulary Practice: Definition Chart
- **GeoActivity** Research Reform Movements

TECHTREK myNGconnect.com

» **Fast Forward!**
Core Content Presentations
Teach *Democracy in Eastern Europe*

Digital Library
NG Photo Gallery, Section 2

Connect to NG
Research Links

Magazine Maker

Also Check Out
- Graphic Organizers in **Teacher Resources**
- GeoJournal in **Student eEdition**

BACKGROUND FOR THE TEACHER

Of all the eastern European countries, Hungary had the smoothest transition from communism to democracy and a market economy. In the 1980s, activists both within and outside the Communist Party began to call for change. In 1988, the parliament adopted a "democracy package" that included such reforms as freedom of assembly and the press, a new electoral law, and a revised constitution.

ESSENTIAL QUESTION

What are the costs and benefits of European unification?

Section 2.3 describes the efforts of eastern European countries to adopt democracy and capitalism and seek membership in the European Union and NATO.

INTRODUCE & ENGAGE

Team Up: Brainstorm Divide the class into groups and give each group a chart like the one at right. Tell the class that their goal is to end the Cold War divisions in Europe. Allow them to brainstorm the obstacles and possible strategies to accomplishing that goal. Allow them to regroup after the class has studied the lesson so they can fill out the Outcomes box. **0:15** minutes

Goal:	End the Cold War divisions in Europe

↓

Obstacles:	*Communist governments and economies*

↓

Strategies:	*Make countries more democratic*

↓

Outcomes:	*Democratization and privatization*

TEACH Digital Library

Guided Discussion

1. Synthesize Review with students the spread of democracy in Eastern Europe and the collapse of the Soviet Union. **ASK:** What was the relationship between the spread of democracy in Eastern Europe and the collapse of the Soviet Union? *(Possible response: The Soviet Union imposed communism on Eastern Europe. The ability of Eastern Europe to overthrow communism showed that the Soviet Union was weakening or perhaps becoming democratic itself. Shortly afterward, the Soviet Union fell apart.)*

2. Explain Why didn't eastern Europeans have freedom of the press and speech under Communist rule? *(Communist governments did not want their citizens to criticize government actions.)*

Analyze Visuals Project the images of Solidarity posters (shown at right) and the Berlin Wall from the **NG Photo Gallery**. Ask students to discuss the messages that the posters convey and the emotions they evoke. Also ask the class to infer how West Germans and East Germans felt when the Berlin Wall came down. **0:10** minutes

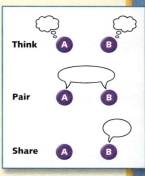

W SAMO POŁUDNIE
4 CZERWCA 1989

DIFFERENTIATE Connect to NG

English Language Learners Use Suffixes Point out that the two Key Vocabulary words in this section end with *-ization*, which means "the process of making or changing something." Explain that this suffix is related to the suffix *-ize*, which means "to cause something to change into something else." The verb *democratize* means to make a system more democratic. The verb *privatize* means to change a business from government ownership to private ownership. Have students practice using the suffixes *-ize* and *-ization* by having them change the word *colony* into two related words.

Pre-AP Create Time Lines Have students use the **Research Links** to find specific events related to the fall of the Berlin Wall and the overthrow of Communist governments in Eastern Europe in 1989. Then have them create an illustrated time line of the significant events from that year.

ACTIVE OPTIONS

Magazine Maker

Have students design and create a magazine on democracy in Eastern Europe today. Divide the class into groups and have each group create one page for the magazine. Some groups can create a page on an individual country and focus on its successes and challenges. Others can create pages on people who have made a difference in the region—both past and present. For example, students may wish to research the reform efforts of Lech Walesa and Vaclav Hável and their importance to the democratic movement in Eastern Europe as well as more current leaders in the region.

Encourage students to include maps, photos, and graphs in their magazine pages. Once all of the pages have been completed, assemble them into a magazine and, as a class, choose its cover. Allow students to read the magazine and use it as a resource. **0:25** minutes

On Your Feet

Think-Pair-Share Have students choose partners. Explain that in each pair, one student will represent an eastern European who wants to return to Communist rule and the other student will represent an eastern European who prefers democracy. Allow the students a few minutes to think about their respective positions. Then each student should interview his or her partner. Finally each pair should present the results of their interviews to the class. **0:20** minutes

Think A B
Pair A B
Share A B

ONGOING ASSESSMENT
READING LAB GeoJournal

ANSWERS

1. Ukraine faced problems with corruption, too much influence on the part of Russia, and anti-democratic rulers. The solution to the problems reflected democratization because the Ukrainians used their vote to oust objectionable leaders.
2. The younger generations have grown up with democracy and have more confidence in their government. They also don't feel their parents' nostalgia for the old ways of doing things.

2.4 Changing Demographics

TECHTREK
myNGconnect.com For an online graph of Europe's changing demographics
Maps and Graphs   Global Issues

Inspiring people to care about the planet
National Geographic Society Mission

Main Idea New immigrants are changing Europe.

Every May, people from Germany, Denmark, Hungary, Bulgaria, and other European countries come together to celebrate Europe's diversity on Europe Day. The celebrations reflect Europe's changing **demographics**, the characteristics or the profile of a human population. The population has become more diverse as people from Africa and Asia have immigrated to Europe.

An Aging Population

For years, Europe has had an **aging population**. In other words, the average age of people on the continent has been rising. This trend has had a number of causes. For one thing, Europeans have been living longer because of better medical care. For another, most families are having fewer children. So today, senior citizens form a higher percentage of Europe's total population.

The trend created a need for more workers to replace the many senior citizens who were retiring. Workers were also needed to keep the economy strong and pay taxes to support such social services as education and health care. As a result, immigrants came to Europe to take the newly available jobs. People immigrated to Great Britain from former colonies, such as India, Pakistan, and Bangladesh. France also attracted immigrants from former colonies, especially Algeria and Morocco. In the 1970s, people from Turkey began coming to Germany for jobs. Germany now has approximately 2 million people of Turkish heritage. The fall of communism in Eastern Europe in the 1980s and 1990s also resulted in increased migration within Europe.

Before You Move On
Summarize In what way did Europe's aging population create a need for immigrants?

It created a need for more workers who could strengthen the economy, pay taxes, and support retirees.

130 CHAPTER 4

KEY VOCABULARY

demographics, n., the characteristics of a human population, such as age, income, and education

aging population, n., a trend that occurs as the average age of a population rises

ACADEMIC VOCABULARY

assimilate, v., to be absorbed into a society's culture

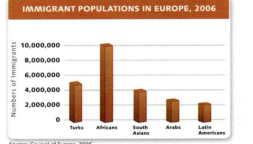

IMMIGRANT POPULATIONS IN EUROPE, 2006

Numbers of Immigrants: Turks, Africans, South Asians, Arabs, Latin Americans (scale 0 to 10,000,000)

Source: Council of Europe, 2006

Reasons for Immigration

Most immigrants have come to Europe to seek a better life. Some left their home countries for economic reasons—usually to find jobs. Others left for political reasons—to escape conflict in their countries or an unjust government. Once they have found work in Europe, many immigrants send money to relatives back home. Eastern Europeans have migrated to Western Europe for similar reasons. Membership within the European Union (EU) has made their migration easier.

Challenges of Immigration

Tensions have sometimes developed between immigrants and native citizens. These tensions often arise when the two groups compete for the same jobs. Also, some immigrants enter European countries illegally—according to some EU estimates, about a half million each year. Housing, educating, and caring for these illegal immigrants can create economic strains on the host countries.

The mix of different cultures has also created problems. Many immigrants are Muslims, with cultural and religious practices that differ from those of their Christian neighbors. Some Europeans would like to see the Muslim immigrants **assimilate**, or be absorbed into their society's culture. They believe the immigrants should adopt European traditions and values. Others believe a more multicultural approach is better. This approach encourages tolerance and embraces all cultures.

Immigrants are bringing their cultural and religious practices, helping to support the economy, and sometimes creating tensions.

COMPARE ACROSS REGIONS

Australia's Skilled Immigrants

Like Europe, Australia has an aging population. Many in the Australian government believe that immigration can help change that trend. As a result, each year the government identifies gaps in the country's workforce. It then determines the number of skilled immigrants that can come to Australia. Between 2008 and 2009, more than 110,000 immigrants arrived on this skilled migration program. Most of these immigrants came from Great Britain, India, and China.

Of course, Europe receives many more immigrants than Australia each year. Italy alone took in more than 400,000 immigrants in 2008. Many Europeans appreciate the cultural enrichment that immigrants bring to their countries. However, some believe that Europe, like Australia, should set immigration limits.

Before You Move On
Monitor Comprehension What are the ways in which new immigrants are changing Europe?

ONGOING ASSESSMENT

READING LAB   GeoJournal

1. **Interpret Graphs** According to the bar graph, from which two continents has the immigrant population in Europe primarily come?
2. **Make Inferences** Why might some immigrants resist assimilation?
3. **Identify Problems and Solutions** What has Australia done to solve some of the problems posed by immigration?

SECTION 2.4 **131**

PLAN

OBJECTIVE Describe the impact of immigration on European people and culture.

CRITICAL THINKING SKILLS FOR SECTION 2.4

- Main Idea
- Summarize
- Monitor Comprehension
- Interpret Graphs
- Make Inferences
- Identify Problems and Solutions
- Make Predictions
- Analyze Causes
- Synthesize

PRINT RESOURCES

Teacher's Edition Resource Bank

- Reading and Note-Taking: Analyze Cause and Effect
- Vocabulary Practice: Related Idea Web
- **GeoActivity** Compare London's Immigrant Populations

TECHTREK   myNGconnect.com

Fast Forward!
Core Content Presentations
Teach *Changing Demographics*

Maps and Graphs
Graph: Immigrant Populations in Europe, 2006

Interactive Whiteboard
GeoActivity Compare London's Immigrant Populations

Also Check Out
- Graphic Organizers in **Teacher Resources**
- GeoJournal in **Student eEdition**

BACKGROUND FOR THE TEACHER

By late 2010, some western Europeans feared that their culture was changing too much because of the high numbers of immigrants. Surveys showed that more than 30 percent of Germans believed the country was "overrun by foreigners," and in October, Chancellor Angela Merkel said in a speech that Germany's approach to building a multicultural society "has failed, utterly failed." In France, a controversial law was passed banning the Muslim practice of women wearing burqas in public.

ESSENTIAL QUESTION

What are the costs and benefits of European unification?

The movement of people from Eastern Europe to Western Europe and to Europe from other parts of the world has created challenges, as discussed in Section 2.4.

INTRODUCE & ENGAGE

Team Up: Make Predictions Have students get into groups and give each group a graphic organizer like the one below. Tell them that they are commissioners for a European country. Their population is getting older, and their country no longer has enough workers. Have them use the chart to predict actions that the country might take to solve this problem. Discuss the graphic organizers after students have completed the lesson. `0:15` minutes

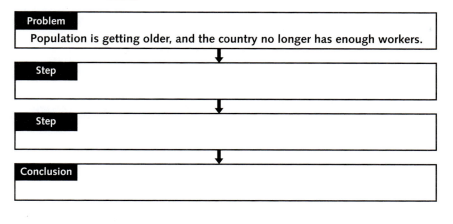

Problem
Population is getting older, and the country no longer has enough workers.

Step

Step

Conclusion

TEACH   Maps and Graphs

Guided Discussion

1. Analyze Causes Why is the age of Europe's population rising? *(People are living longer because of better medical care, and most families are having fewer children.)*

2. Synthesize How has membership in the European Union made it easier for eastern Europeans to migrate to Western Europe? *(Citizens of EU countries can travel freely anywhere in the EU.)*

Interpret Graphs Project the Immigrant Populations in Europe graph. Point out that Arabs and South Asians both come from Asia, as do most people from Turkey. **ASK:** How does the total number of Asian immigrants compare to the number of African immigrants? *(It is about the same.)* Why do you think there is no bar for North America? *(The United States and Canada have prosperous economies and democratic governments, so there is little incentive for large groups of people to emigrate from there.)* `0:15` minutes

DIFFERENTIATE

Striving Readers **Summarize** Have students work in pairs. Assign each pair one of the four sections under the blue subheadings. Pairs should read their sections together and then write a summary sentence to share with the class. If more than one pair summarizes a particular section, allow them to take turns sharing their section summary before moving to the next section.

Gifted & Talented **Discuss Reasons for Immigration** Have groups of students discuss why immigrants come to Europe. Encourage them to explain both the push factors that might have caused people to leave their home countries and the pull factors that drew them to Europe.

ACTIVE OPTIONS

Interactive Whiteboard
GeoActivity

Compare London's Immigrant Populations Have students work in pairs to read the passage aloud and complete the chart of push-pull factors fueling immigration to London. Then have them answer the critical thinking questions. `0:15` minutes

On Your Feet

Fishbowl Have half the class sit in a close circle, facing inward. The other half of the class sits in a larger circle around them. Post the question *How is Europe's population changing?* Students in the inner circle should discuss the question for ten minutes while those in the outer circle listen to the discussion and evaluate the points made. Then have the groups reverse roles and continue the discussion. `0:25` minutes

Performance Assessment

Conduct a Panel Discussion Have groups of students form a panel to discuss government and economics in Europe. Students should talk about the benefits and challenges of each. They should also propose one step European governments can take to resolve issues. Have students use information from the section to support their ideas. Go to **myNGconnect.com** for the rubric.

ONGOING ASSESSMENT
READING LAB   GeoJournal

ANSWERS
1. Asia and Africa
2. They might not want to give up their religion and culture.
3. Australia has a program that allows in a fixed number of skilled immigrant workers based on the country's needs.

VOCABULARY

Match each word in the first column with its meaning in the second column.

WORD	DEFINITION
1. abstract	a. tax on imports and exports
2. dialect	b. control over one's own affairs
3. sovereignty	c. art style emphasizing form and color
4. tariff	d. absorb into another culture
5. privatization	e. privately owned businesses
6. assimilate	f. regional language

MAIN IDEAS

7. Why are there so many different dialects in Europe? (Section 1.1)

8. What is the fastest-growing religion in Europe? Why might that be so? (Section 1.1)

9. During the Romantic period, what themes did artists use in their paintings? (Section 1.2)

10. Why do you think the ancient Greeks and Romans celebrated historic events in epic poems? (Section 1.3)

11. How does the soup called *borscht* reflect Russia's geography? (Section 1.4)

12. What is one of the requirements for joining the European Union? (Section 2.1)

13. In what way has the euro helped increase trade among European nations? (Section 2.2)

14. Why did the Ukrainian people stage the Orange Revolution? (Section 2.3)

15. What factors have led people from other parts of the world to immigrate to Europe? (Section 2.4)

CULTURE

ANALYZE THE ESSENTIAL QUESTION

How is the diversity of Europe reflected in its cultural achievements?

Critical Thinking: Draw Conclusions

16. In what ways do the many dialects in Europe show its great diversity?

17. What aspects of ancient Greek and Roman art inspired Renaissance artists?

18. Why did pasta with tomato sauce develop in Italy rather than Russia?

INTERPRET MAPS

CHRISTIANITY IN EUROPE

Protestant
Roman Catholic
Eastern Orthodox
Other religions

19. **Region** Where do most Protestants live in Europe? Where do most Catholics live?

20. **Make Inferences** Find the countries on the map in which a relatively small part of the population belongs to a different Christian denomination. What challenges might the people in the minority religion face?

GOVERNMENT & ECONOMICS

ANALYZE THE ESSENTIAL QUESTION

What are the costs and benefits of European unification?

Critical Thinking: Analyze Cause and Effect

21. In what way did the Marshall Plan help bring about the formation of the European Union?

22. What was the response of the eurozone countries when Greece and Ireland became deeply in debt in 2010?

23. What was the impact of the fall of communism on Eastern Europe? What was the impact on Western Europe?

24. What problems are caused by illegal immigration to Europe?

INTERPRET CHARTS

COST OF A TEN-MINUTE PHONE CALL TO THE U.S. IN EUROS (€)*		
Country	1997	2006
Belgium	7.50	1.98
Czech Republic	3.09	2.02
Denmark	7.41	2.38
Ireland	4.61	1.91
Spain	6.17	1.53
France	6.78	2.32
United Kingdom	3.50	2.23

* 1997 prices have been converted to euros
Source: Eurostat

25. **Analyze Data** According to the chart, how did the cost of making a telephone call change between 1997 and 2006?

26. **Analyze Cause and Effect** What move made by the European Union might have brought about the change in the cost of a telephone call?

ACTIVE OPTIONS

Synthesize the Essential Questions by completing the activities below.

27. **Write a Speech** Suppose that you are the leader of a European country that has been invited to join the European Union. Write a speech that will persuade the citizens of your country to vote in favor of joining. **Deliver your speech to the class and ask the members of your audience to vote on whether they are in favor of joining the European Union.**

> **Writing Tips**
> - Take notes on three benefits that would result from joining the European Union.
> - Support each benefit using facts and statistics.
> - Address any concerns your audience might have about joining and explain why the advantages outweigh any disadvantages.

TECHTREK myNGconnect.com For photos of European art

28. **Create an Art Chart** Select three European works of art. You can choose Raphael's *School of Athens* or Leonardo da Vinci's *Mona Lisa* from the **Digital Library**, or you can search for other works online. Then research each piece to find out what period it is from and what theme, or subject matter, it represents. Copy the chart below to help you organize your information. Display the artwork and your findings on a poster. Be prepared to explain the relationship between each work's period and theme.

WORK OF ART	PERIOD	THEME

CHAPTER Review

VOCABULARY ANSWERS

1. c
2. f
3. b
4. a
5. e
6. d

MAIN IDEAS ANSWERS

7. There are more than 40 countries, creating a great variety of languages, and even within countries, people in separate regions have developed their own variations of the national language.

8. Islam is the fastest-growing religion, perhaps because of immigration from other countries.

9. They painted landscapes and other scenes to convey emotion.

10. Possible response: They wanted to celebrate the heroes that were important to their cultures.

11. The soup is made of beets, a crop that is well adapted to Russia's climate.

12. having a stable democracy that respects human rights

13. It has lowered transaction fees and made costs and prices easier to compare.

14. to remove their prime minister, Viktor Yanukovych, from power

15. for economic reasons or to escape conflict or an unjust government in their countries

CULTURE

ANALYZE THE ESSENTIAL QUESTION ANSWERS

16. Dialects reflect the many ethnic groups and cultures within Europe.

17. Like the ancient artists, Renaissance artists presented human subjects realistically.

18. Some of the ingredients do not grow well in Russia.

INTERPRET MAPS

19. Northern Europe; Southern Europe

20. They might face discrimination.

GOVERNMENT & ECONOMICS

ANALYZE THE ESSENTIAL QUESTION ANSWERS

21. To manage the U.S. aid money, European countries formed the Organization for European Economic Cooperation and discovered they could accomplish more by working together.

22. The eurozone loaned money to Greece and Ireland to help the countries out of their debt crisis.

23. After the fall of communism, Eastern European countries developed democratic governments and market economies. They traded much more with Western Europe. Several eastern European countries have joined the European Union.

24. It has created economic strains on the host countries.

INTERPRET CHARTS

25. The cost of the calls decreased.

26. the creation of the euro

ACTIVE OPTIONS

WRITE A SPEECH

27. Speeches should
- cite three benefits of joining the European Union;
- support each benefit with facts and statistics;
- address counterarguments with explanations about why the advantages outweigh the disadvantages;
- be delivered in a strong voice with clear enunciation.

CREATE AN ART CHART

28. The chart should present three works of art and correctly identify the period and theme of each one. Students' choices will vary, but their charts should clearly show they have researched the pieces and organized their information effectively. Sample responses for the chart are at right.

WORK OF ART	PERIOD	THEME
Leonardo da Vinci's *Mona Lisa*	16th century Renaissance	The Mona Lisa shows a synthesis of various styles of portraiture and the figure in the landscape.
Raphael's *School of Athens*	16th century Renaissance	The School of Athens celebrates classical Greek philosophy, science, and art.
Michelangelo's *David*	16th century Renaissance	David celebrates the human form in this realistic portrayal.

World Languages

TECHTREK
myNGconnect.com For an online graph
and research links on world languages

Maps and Graphs | Connect to NG

In this unit, you learned about the diversity of languages in Europe. Every culture in the world uses language to communicate. Scholars estimate that there are about 7,000 languages spoken today.

Most countries name one or more languages as official languages. An official language is the one used by a country's government. For example, French is the official language of France, and English and French are the official languages of Canada. Most countries have groups of people whose first language differs from the official language. It is estimated that at least half of the people in the world speak one or more languages in addition to their first language.

Compare

- Africa
- Americas
- Asia
- Europe
- Oceania

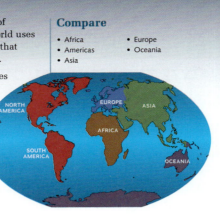

LIVING LANGUAGES

Although there are thousands of world languages, many are not widely spoken. In addition, sometimes the distinction, or difference, between a language and a dialect is not clear.

The following are the ten most widely spoken languages in the world in order of their ranking. Each is spoken as a native language by at least 100 million people. Some are official languages in widely different regions of the world. Some of the languages appear in only one region of the world.

1. Mandarin Chinese
2. Spanish
3. English
4. Hindi/Urdu
5. Arabic
6. Bengali
7. Portuguese
8. Russian
9. Japanese
10. German

DYING LANGUAGES

Some languages are spoken by so few people that they are in danger of dying out. In fact, linguists, or people who study languages, estimate that one language dies every two weeks.

Languages die for different reasons. Some simply disappear with the death of the last speaker. Others fade more slowly as a dominant language replaces it. A few linguists are trying to document some of these languages to preserve the history and culture of the people who spoke them.

The graph at right shows the number of languages spoken on the world's continents, as well the names of a few of the languages spoken. Note that "Americas" consists of the number of languages spoken on the North American and South American continents. Compare the data in the graph and use it to answer the questions.

NUMBER OF LANGUAGES SPOKEN BY CONTINENT

Source: *Ethnologue*, 16th Edition, 2009

- EUROPE
- AFRICA
- ASIA
- AMERICAS
- OCEANIA

ONGOING ASSESSMENT
RESEARCH LAB — GeoJournal

1. **Compare and Contrast** On which continent are the most languages spoken? the fewest? What do these numbers suggest about the cultural unity of each continent?

2. **Analyze Data** Study the graph and the list of languages with the most native speakers. How many of the total languages in Europe are among those spoken most in the world?

Research and Create Charts Research to find out more about the people who speak Hindi and Portuguese. Create a chart for each language showing approximately how many people speak it and where it is spoken. Which language has more native speakers? Which language is an official language in more places? What might account for this?

PLAN

OBJECTIVE Compare languages using data from graphs.

CRITICAL THINKING SKILLS

- Compare and Contrast
- Analyze Data
- Create Charts
- Draw Conclusions
- Interpret Graphs

PRINT RESOURCES

Student Edition Economics and Government Handbook

Teacher's Edition Resource Bank

Review: GeoActivity Analyze the Roots of Modern Languages

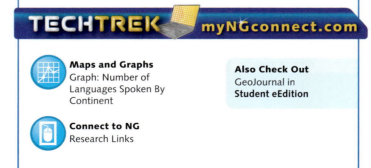

TECHTREK myNGconnect.com

Maps and Graphs
Graph: Number of Languages Spoken By Continent

Connect to NG
Research Links

Also Check Out
GeoJournal in **Student eEdition**

BACKGROUND FOR THE TEACHER

According to National Geographic's Enduring Voices Project, about 3,500 languages are endangered and at risk of becoming extinct. The criteria that language experts use to determine if a language is endangered is the percentage of young people and children who are able to learn and speak it. If less than 30 percent of this population falls into that category, then a language is endangered. In addition, some governments discourage the use of lesser-spoken, "minor" languages, which often leads to their extinction.

In the late 20th century, advocates for language preservation began to organize and become more vocal for change. To raise awareness of endangered languages, International Mother Language Day was established on February 21, 2000. The United Nations Educational, Scientific, and Cultural Organization (UNESCO) has set aside this day each year to promote world language recognition in hopes of preserving existing and endangered languages.

INTRODUCE & ENGAGE

Hands-On Geography Have students form small groups, and provide each one with an outline map of a different country. Have students research to identify their country's official language (or languages) and three to five other languages that are spoken there. Ask students to label these languages on their map, using a legend to identify whether a language is official or not. Once the groups have finished their work, display the maps. **ASK:**

- In which countries are the same languages spoken? *(Responses will vary.)*
- Which countries have more than one official language? Why do you think that might be? *(Responses will vary, but students should say a country might have more than one language if a significant percentage of the population speaks it.)*
- What factors might influence what languages are spoken in a country? *(the people who first settled there, conquerors, immigrants)* `0:15` **minutes**

TEACH · Maps and Graphs

Guided Discussion

1. **Compare and Contrast** What is the difference between native languages and official languages? *(Native languages are spoken by native speakers in a country or region while official languages are those that are used by the country's government.)*

2. **Draw Conclusions** Why do linguists want to try to preserve dying languages? *(They want to preserve a people's history and culture, and language is a large part of those. If the language dies out, the history and culture of the people are affected and may not be passed on to future generations.)*

Interpret Graphs Point out some of the languages on the graph that appear on several continents. **ASK:** What languages that are common in the Americas and Europe surprised you? *(Possible response: Dutch)* in Asia and the Americas? *(Possible response: Hindi)* in Africa and Oceania? *(Possible response: French)* in Europe and Asia? *(Possible response: Greek)* in Africa and Asia? *(Possible response: Arabic)* in Africa and Europe? *(Possible response: Spanish)*

Have students draw a web for each language that appears in the graph on more than one continent. **ASK:** Which continent appears in more webs than the others? *(Possible response: Africa)* `0:20` **minutes**

Spanish

DIFFERENTIATE

English Language Learners Interpret Graphs Point out that some languages appear more than once in the bar graph. Explain that a continent includes many countries and that often there are several languages spoken in a country even though there may only be one or two official languages. Have pairs work together to interpret the data.

On Your Feet

Reinterpret Data After students have completed the Research Lab, have them form groups of five. Have each group compare the data for the five continents and choose the continent that seems to have the most diversity in languages. Students should note the languages that are spoken on more than one continent, which will give them a better idea of which continent has the most diversity.

Have a volunteer from each group write the name of the continent they chose on the board. If groups have identified different continents, ask them to share their tally count, and then discuss their choices. `0:15` **minutes**

Performance Assessment

One Hundred Million Speakers Arrange the class into pairs or groups of three and assign each pair or group one of the world's ten most widely spoken languages listed in the text. Ask each pair or group to find the language on the graph in the text and to find out on how many continents the language is spoken. Students may use the **Research Links** to verify where their assigned language is spoken and how many speakers of the language there are on each continent. You might have students use a chart like the one below to keep track of their information. Ask pairs or groups to share their results with the class. Go to **myNGconnect.com** for the rubric.

CONTINENT	NUMBER OF SPEAKERS
Africa	
Americas	
Asia	
Europe	
Oceania	

RESEARCH LAB · GeoJournal

ANSWERS

1. Asia; Europe; The numbers suggest that Europe is more culturally unified than Asia.
2. five: Spanish, English, Portuguese, Russian, German

Research and Create Charts Hindi; Portuguese; Portugal probably spread its language through its voyages of exploration in the 1400s and the colonies it established throughout the world.

Active Options

TECHTREK
myNGconnect.com For photos of Renaissance art, nuclear power plants in Europe, and European cuisine

Digital Library · Connect to NG · Magazine Maker

ACTIVITY 1

Goal: Extend your understanding of Renaissance art.

Write a Renaissance Arts Magazine

The Renaissance was a period of great artistic activity in Europe. Choose a city in Europe that was influenced by the Renaissance between 1400 and 1600. With a group, plan and publish a magazine showcasing that city's artistic achievements. Use the Magazine Maker to find photos and information on the following:

- art
- architecture
- literature
- fashion

Brunelleschi's dome atop the Cathedral of Florence in Italy

ACTIVITY 2

Goal: Research the use of nuclear power in Europe.

Create a Pro-and-Con Chart

Some European countries are planning to build new plants, while others have chosen to close existing plants. Use the research links at **Connect to NG** to create a pro-and-con chart that explains some of the advantages and disadvantages of nuclear power. Be prepared to present your chart and explain the issues.

ACTIVITY 3

Goal: Learn about European culture through its food.

Plan a Dinner Menu

Get together in a group and plan a dinner menu featuring typical European dishes. Discuss what the courses for the meal will be. Each group member should be in charge of one course, each of which should come from a different European country. Design a poster presentation of the menu.

ASSESS

Use the rubrics to assess each student's participation and performance.

Project Rubric: Activity 1

SCORE	Planning / Preparation	Content / Presentation	Participation / Collaboration
3 GREAT	• Conducts appropriate research to gather information and art. • Prepares an excellent plan to help with the organization of the magazine.	• Incorporates strong artistic achievements in all areas. • Incorporates many supporting visual aids. • Appeals to the audience.	• Participates in group work for the magazine publication. • Takes a leading role in planning the magazine publication.
2 GOOD	• Conducts some research to gather information and art. • Prepares a good plan to help with the organization of the magazine.	• Incorporates some artistic achievements in some areas. • Incorporates a few supporting visual aids. • Moderately appeals to the audience.	• Participates somewhat in group work for the magazine publication. • Takes a subordinate role in planning the magazine publication.
1 NEEDS WORK	• Conducts little to no research to gather information and art. • Does not prepare a plan to help with the organization of the magazine.	• Incorporates few or no artistic achievements. • Does not incorporate supporting visual aids. • Does not appeal to the audience.	• Does not participate in group work for the magazine publication. • Does not take any role in planning the magazine publication.

Project Rubric: Activity 2

SCORE	Planning / Preparation	Content / Presentation	Participation / Collaboration
3 GREAT	• Conducts appropriate research to gather information. • Prepares notes to help with organization and presentation.	• Includes multiple accurate advantages of nuclear power. • Includes multiple accurate disadvantages of nuclear power.	• Presents clear arguments for and against the use of nuclear power. • Provides thorough explanations for data in chart.
2 GOOD	• Conducts some research to gather information. • Prepares some notes to help with organization and presentation.	• Includes some accurate advantages of nuclear power. • Includes some accurate disadvantages of nuclear power.	• Presents somewhat clear arguments for and against the use of nuclear power. • Provides some explanations for data in chart.
1 NEEDS WORK	• Conducts little to no research to gather information. • Does not prepare notes to help with organization and presentation.	• Includes few to no advantages of nuclear power. • Includes few to no disadvantages of nuclear power.	• Does not present clear arguments for and against the use of nuclear power. • Provides few to no explanations for data in chart.

Project Rubric: Activity 3

SCORE	Planning / Preparation	Content / Presentation	Participation / Collaboration
3 GREAT	• Reviews several sources to identify common dishes. • Conducts appropriate research to gather information.	• Poster is visual, accurate, and complete. • Visuals include excellent captions, and menu is representative of European dishes.	• Works well with others. • Assumes a clear role and related responsibilities.
2 GOOD	• Reviews a few sources to identify common dishes. • Conducts some research to gather information.	• Poster is somewhat visual and mostly accurate. • Visuals include clear captions, and menu is somewhat representative of European dishes.	• Works well with others most of the time. • Sometimes has difficulty sharing decisions and responsibilities.
1 NEEDS WORK	• Does not review sources to identify common dishes. • Conducts little to no research to gather information.	• Poster is not visual or accurate. • Visuals do not include captions, and menu is not representative of European dishes.	• Cannot work with others in most situations. • Cannot share decisions or responsibilities.

RESOURCE BANK
EUROPE

EUROPE RESOURCE BANK

GEOGRAPHY & HISTORY

RESOURCE BANK

SECTION 1 GEOGRAPHY

1.1 Physical Geography

Use with Europe Geography & History, Section 1.1, *in your textbook.*

Reading and Note-Taking Create a Sketch Map

As you read about Europe's physical geography in Section 1.1, sketch your own physical map of Europe. Draw an outline map of the continent. Include islands such as Great Britain and Iceland. Then add labels according to the directions below.

1. What are Europe's six climate regions?

☐ *Semiarid* _____   ☐ _____

☐ _____   ☐ _____

☐ _____   ☐ _____

2. Create a map key for the six climate regions by shading in the rectangles next to each region you have listed above. Use a different shade for each region. Shade in your map to show the location of each region.

3. Write down the four land regions that form Europe. Then label those regions on the map.

_____   _____

_____   _____

4. Label the Italian, Scandinavian, and Iberian peninsulas.

5. Label the Atlantic Ocean, Mediterranean Sea, Black Sea, and the North Sea.

© NGSP & HB

SECTION **1** GEOGRAPHY

1.1 Physical Geography

Use with Europe Geography & History, Section 1.1, *in your textbook.*

Vocabulary Practice

KEY VOCABULARY

- **peninsula** (puh-NIHN-suh-lah) n., a piece of land that is nearly surrounded by water, but is attached to a larger land area by a thin strip of land

- **uplands** n., an area of high land along rivers or between hills

Descriptive Paragraph Write a paragraph describing the physical geography of Europe. In your description, define the Key Vocabulary words *peninsula* and *uplands* and explain where these landforms are found on the continent.

_____

_____

_____

_____

_____

_____

_____

_____

_____

_____

_____

_____

_____

Name _____ Class _____ Date _____

GeoActivity

Use with Europe Geography & History, Section 1.1, in your textbook.

Go to Interactive Whiteboard GeoActivities at
myNGconnect.com to complete this activity online.

**NATIONAL
GEOGRAPHIC**
School Publishing

1.1 PHYSICAL GEOGRAPHY

Map Europe's Land Regions

Study the chart below to learn more about Europe's land regions. Then use
the information to shade in the regions on the map and answer the questions.

Europe's Land Regions and Economy

LAND REGION	LOCATIONS	TRAITS	ECONOMIC ACTIVITY
Western Uplands	Scandinavian Peninsula, Scotland, Ireland, Iceland, Portugal, Spain	Rugged highlands, poor soil, many marshlands and lakes	Livestock, hydroelectric power
Northern European Plain	England, northern France and Germany, northeastern Europe	Good soil, river systems	Farming, trade
Central Uplands	Eastern France, Luxembourg, southern Germany	Hills, forests, rough plateaus	Livestock, mining
Alpine Region	Mountains of Spain, Switzerland, Italy, Greece, and southeastern Europe	Mountains	Vineyards, orchards, olive groves

1. **Create Maps** Use the information above to plot the four land regions of
Europe on the map. Choose a color for each region and shade in the map.
Remember to create a legend for the map.

2. **Make Generalizations** How might the geographic features of
a region affect its economy? Choose one land region and explain.

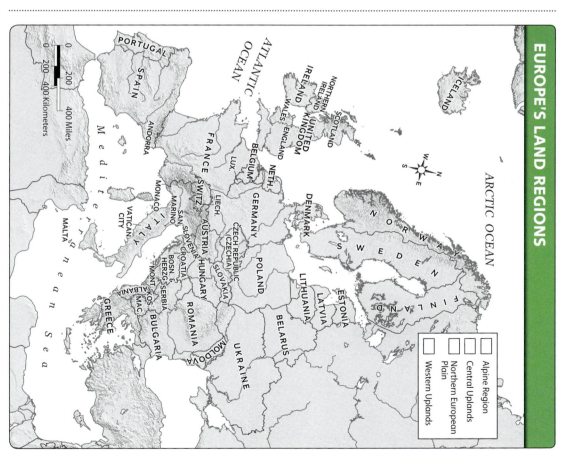

EUROPE'S LAND REGIONS

Legend:
- Alpine Region
- Central Uplands
- Northern European Plain
- Western Uplands

© NGSP & HB

Resource Bank **UNIT 2** **RB7**

SECTION **1** GEOGRAPHY

1.2 A Long Coastline

Use with Europe Geography & History, Section 1.2, *in your textbook.*

Reading and Note-Taking Analyze Cause and Effect

Use a Cause and Effect Organizer as you read Section 1.2 to analyze the effects of Europe's long coastline on trade, industry, exploration, and settlement.

Cause
Europe has more than 24,000 miles of coastline.

Effect 1 **Trade:** Having so much water access helped the growth of trade, which was central to the growth of the civilizations of ancient Greece and Rome.

Effect 2 Industry:

Effect 3 Exploration:

Effect 4 Settlement:

© NGSP & HB

Name   Class   Date

Use with Europe Geography & History, Section 1.2, *in your textbook.*

SECTION **1** GEOGRAPHY
1.2 A Long Coastline

Vocabulary Practice

KEY VOCABULARY

- **bay** n., a small body of water set off from the main body of water
- **fjord** (fee-ORD) n., a narrow inlet of the sea between cliffs, steep hills, or mountains
- **polder** n., land that has been reclaimed, or taken, from the seabed

Definition Chart Complete a Definition Chart for the Key Vocabulary words. Write a sentence using each word as it relates to what you learned in Section 1.2.

Word	bay	fjord	polder
Definition	a small body of water set off from the main body of water		
In Your Own Words			
Sentence			

© NGSP & HB

Name _____ Class _____ Date _____

Use with Europe Geography & History, Section 1.2, in your textbook.

Go to Interactive Whiteboard GeoActivities at
myNGconnect.com *to complete this activity online.*

SECTION 1 GEOGRAPHY
GeoActivity

1.2 A LONG COASTLINE

Analyze Early European Trade

Read the passage below to learn more about trade routes at the time of the Roman Empire. Then answer the questions.

Trade in the Roman Empire

At the height of its power, the Roman Empire dominated the entire Mediterranean region, most of western Europe, and large areas of northern Africa. Romans traded by sea routes and a vast network of roads. Grain and marble were imported from North Africa. Wine, olive oil, and honey came by ship from Spain, as well as gold and silver from mines there. Cloth and glassware were imported from Syria. The Romans traded gold and gems for silks from China. Spices came from South Asia. Timber used for building ships came from Macedonia. Metals used in making weapons were mined in Britain, and wool was also imported from the island. Those goods were exchanged for pottery, wine, and olive oil brought in by the Romans.

Towns and cities in many parts of the empire built places—small "factories"—to make products for trading. People were also needed to unload or load goods, count them, weigh them, and pack them again in order to be transported.

1. **Create Maps** Using the information in the passage, create and place symbols for different goods on the map. Identify your symbols in the map's legend.

2. **Interpret Maps** Choose one good that you placed on the map. Use a highlighter to mark the different trade routes that might carry that good to Rome. What other cities might the good pass through?

3. **Make Inferences** Notice how many cities on the map are close to major bodies of water. How does geographic location influence and affect trade?

TRADE ROUTES OF THE ROMAN EMPIRE

Legend:
- Roman Empire at its height, c. 200
- Principal trade routes by sea
- Principal trade routes by land

© NGSP & HB

SECTION **1** GEOGRAPHY
1.3 Mountains, Rivers, and Plains

Use with Europe Geography & History, Section 1.3, *in your textbook.*

Reading and Note-Taking Pose and Answer Questions

Use the KWL Chart below to record what you already know about Europe's physical features and then pose questions in the W column about what you want to find out. Finally answer your questions in the L column.

K What Do I Know	W What Do I Want To Learn?	L What Did I Learn?
Europe has a variety of landforms and natural resources.		

SECTION **1** GEOGRAPHY

1.3 Mountains, Rivers, and Plains

Use with Europe Geography & History,
Section 1.3, *in your textbook.*

Vocabulary Practice

KEY VOCABULARY

- **canal** (kah-NAL) n., a long narrow waterway created by people for travel, shipping, or irrigation
- **waterway** n., a canal or river that is navigable

ACADEMIC VOCABULARY

- **navigable** (NAV-ih-gah-buhl) adj., deep and wide enough for boats and ships to travel on or through

Write a sentence using the Academic Vocabulary word *navigable*.

_____

_____

Venn Diagram Complete the Venn Diagram below for the Key Vocabulary words *canal* and *waterway*. Next to each word, write its definition in your own words. Write down what they have in common in the middle space where the two circles intersect.

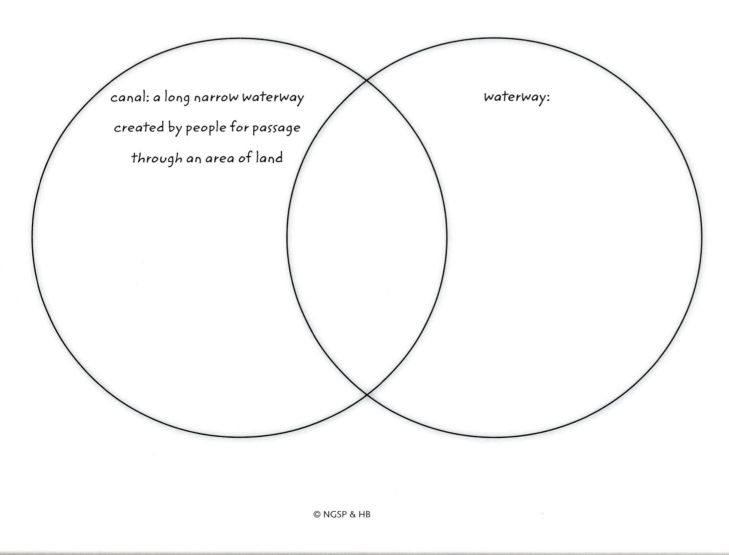

canal: a long narrow waterway created by people for passage through an area of land

waterway:

☐ NATIONAL GEOGRAPHIC School Publishing

SECTION ❶ GEOGRAPHY

GeoActivity

Use with Europe Geography & History, Section 1.3, in your textbook.

Go to Interactive Whiteboard GeoActivities at **myNGconnect.com** to complete this activity online.

1.3 MOUNTAINS, RIVERS, AND PLAINS

Explore Amsterdam's Canals

Read the passage below to learn more about Amsterdam in the Netherlands and the role the canal system plays in city life. Then use the information in the passage and the map to answer the questions.

Amsterdam's Canals

Water control played a major role in the development of the city of Amsterdam. The city and its port are located in the western Netherlands. Before the 12th century, the area was barely habitable—it was mostly bogs and wetlands at the end of the Amstel River, which tended to flood. Villagers built a dam to control the floods and to create a harbor for ships and trade. The Amstel Dam is how the city got its name.

The first canals were dug in the early 1400s as the city expanded. Canals provided protection as well as access to the sea. They created polders, or artificially drained land, on which buildings could be constructed. The city expanded in the 1500s, and by the 1600s, trade exploded. To keep up with the rapid growth, more canals were built around the old city center. The maze of connecting canals could bring in ships to more merchants and warehouses. In time, hundreds of streets and smaller canals fanned out from the center, creating 90 islands and 1,280 bridges—a marvel of city planning. Amsterdam today has about 62 miles of canals.

1. Identify Problems and Solutions What major problem did the earliest settlers of Amsterdam encounter? How did they solve the problem?

_____

_____

THE CANALS OF AMSTERDAM

Map Tip The suffix -gracht means "canal" in Dutch.

Singelgracht

Prinsengracht
Keizersgracht
Herengracht

Brouwersgracht

National Museum for Art & History

Kloveniers-burgwal

Zwanen-burgwal

Oude Kerk

Central Station

City Hall & Opera

Amstel River

University of Amsterdam

IJ

Brantasgracht
Lamonggracht
Majanggracht
Seranggracht

IJ Bay

IJ

⚓ Dock

0   500   1,000 Meters
0   1,500   3,000 Feet

North Sea

NETHERLANDS
Amsterdam
BELGIUM
GERMANY

2. Analyze Cause and Effect How did the Netherlands' geography affect Amsterdam's economy?

_____

_____

3. Interpret Maps What pattern do you notice in the layout of Amsterdam's canals? Draw additional canals on the map to add to the pattern. Why might city planners have created this pattern?

_____

_____

SECTION **1** GEOGRAPHY

Use with Europe Geography & History, Section 1.4, *in your textbook.*

1.4 **Protecting the Mediterranean**

Reading and Note-Taking **Identify Problems and Solutions**

As you read about Enric Sala's work protecting the Mediterranean Sea in Section 1.4, use the Problem and Solution Diagram below to help you identify problems and solutions.

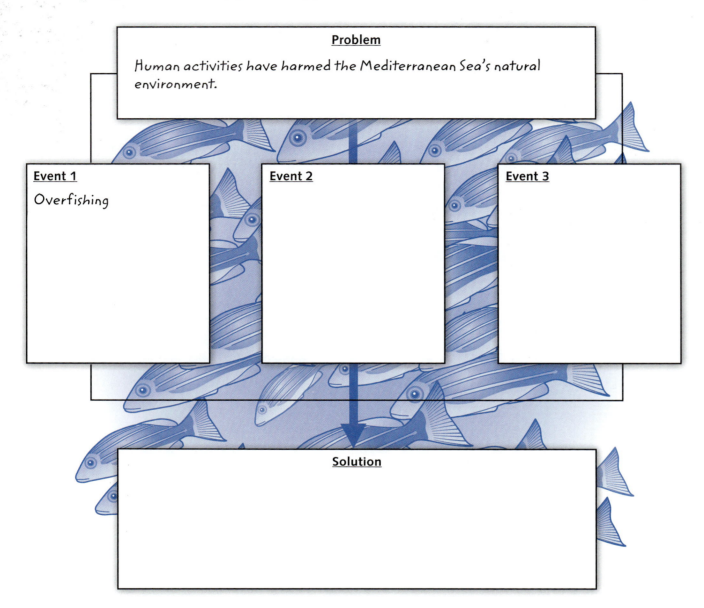

Problem

Human activities have harmed the Mediterranean Sea's natural environment.

Event 1

Overfishing

Event 2

Event 3

Solution

Make a Prediction What do you think would be the result of having more marine reserves in the Mediterranean Sea?

_____

_____

_____

SECTION **1** GEOGRAPHY

1.4 Protecting the Mediterranean

Use with Europe Geography & History, Section 1.4, *in your textbook.*

Vocabulary Practice

KEY VOCABULARY

- **ecosystem** (EE-koh-sis-tehm) n., a community of living organisms and their natural environment
- **marine reserve** n., an area of the sea where animals and plants are given special protection

ACADEMIC VOCABULARY

- **erosion** (ee-ROH-zhuhn) n., the process by which rocks and soil slowly break apart and are swept away

Use the Academic Vocabulary word *erosion* in a sentence about the Mediterranean coastline.

_____

_____

Comparison Chart Complete the Y-Chart to compare the meanings of the Key Vocabulary words *ecosystem* and *marine reserve*. Then write how the two words are related.

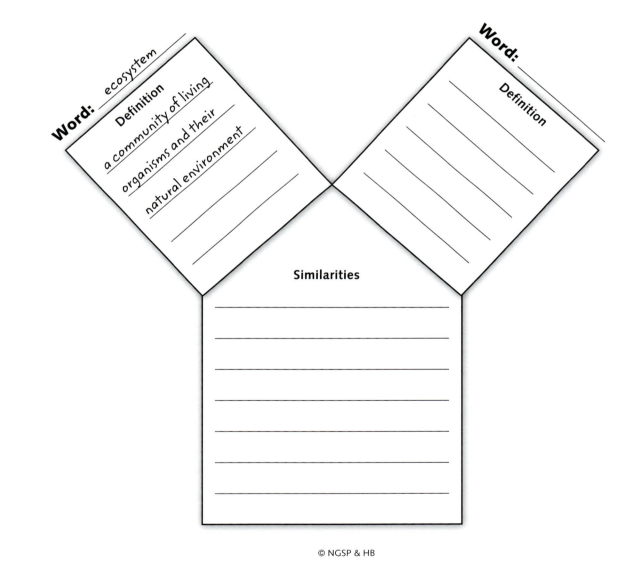

Word: ecosystem
Definition
a community of living organisms and their natural environment

Word: _____
Definition

Similarities

Name _____   Class _____   Date _____

☐ NATIONAL GEOGRAPHIC
School Publishing

SECTION **1** GEOGRAPHY
GeoActivity

Use with Europe Geography & History, Section 1.4, in your textbook.

Go to Interactive Whiteboard GeoActivities at
myNGconnect.com *to complete this activity online.*

1.4 PROTECTING THE MEDITERRANEAN

Graph Fishery Catches in the Mediterranean

Since ancient times, fishing has been a major way of life for people who live around the Mediterranean Sea. Human activities in this region have had serious effects on many kinds of fish and other sea creatures, especially those that people like to eat. As a result, some species have been overfished. Pollution has also affected the ecosystem, while population growth and development have made these problems more intense.

Use the data in the chart to create a line graph that shows how fishery catches have changed for two Mediterranean countries, Italy and Greece. Then answer the questions.

Annual Fishery Catch
(thousands of metric tons)

YEAR	GREECE	ITALY
1950	50	190
1960	85	220
1970	90	400
1980	110	430
1990	130	350
2000	100	300
2005	90	300

Source: Food and Agriculture Organization of the United Nations

1. **Create Graphs** Use the figures in the table to make a double line graph. Use a solid line for Greece and a dotted line for Italy. Label each line on the graph.

Annual Fishery Catch (1950–2005)

ANNUAL FISHERY CATCH
(thousands of metric tons)

500
450
400
350
300
250
200
150
100
50
0

1950 1960 1970 1980 1990 2000 2005

DECADES

2. **Analyze Data** What was the general trend in fishery catches from 1950–1980? How did catches for Greece and Italy compare?

_____

3. **Make Inferences** Did fishery catches for these two countries increase or decrease between 1990 and 2005? What reasons can you suggest for this change?

_____

© NGSP & HB

SECTION **1** GEOGRAPHY
1.1–1.4 Review and Assessment

Use with Europe Geography & History, Sections 1.1–1.4, *in your textbook.*

Follow the instructions below to review what you have learned in this section.

Vocabulary Next to each vocabulary word, write the letter of the correct definition.

1.____ peninsula
2.____ polder
3.____ bay
4.____ fjord
5.____ canal
6.____ ecosystem
7.____ marine reserve

A. a long narrow waterway created by people for travel, shipping, or irrigation
B. a community of living organisms and their natural environment
C. an area of the sea where animals and plants are given special protection
D. a piece of land that is nearly surrounded by water but is attached to a larger land area by a thin strip of land
E. a narrow inlet of the sea between cliffs, steep hills, or mountains
F. land that has been reclaimed, or taken, from the seabed
G. a small body of water set off from the main body of water

Main Ideas Use what you've learned about the geography of Europe to answer these questions.

8. **Identify** What are the four land regions that form Europe?

9. **Describe** How do ocean currents and winds affect climate in Europe?

10. **Contrast** How is a Mediterranean climate different from the climate in Eastern Europe and in Northern Europe?

11. **Main Idea and Details** How has Europe's water access benefited the continent?

12. **Analyze** Why do you think villages in Europe developed near bays?

13. **Identify** What are the natural resources provided by the mountain chains in the Alpine region?

14. **Draw Conclusions** Why do you think some of the most heavily populated cities are located in the Northern European Plain?

15. **Explain** What two human activities have had a negative impact on the Mediterranean's ecosystem?

© NGSP & HB

Focus Skill: Evaluate

Answer the questions below to evaluate how Europe's physical geography encouraged interaction with other regions.

16. What important characteristics of Europe's physical geography helped the rise of industries?

_____

_____

17. How do you think early sailors helped to influence European cultures?

_____

_____

18. How did building polders in the Netherlands change the landscape and economy of the region?

_____

_____

19. What difference have major navigable rivers made to Europe?

_____

_____

20. In what way do Europe's mountain chains help support trade?

_____

_____

21. How has the building of canals supported economic activities in Europe?

_____

_____

22. Why might rivers and canals be better than land for transporting goods?

_____

_____

23. How has a growing population harmed the Mediterranean Sea's physical geography and ecosystem?

_____

_____

Synthesize: Answer the Essential Question

How did Europe's physical geography encourage interaction with other regions? Consider information you have learned about European civilizations, Europe's coastlines and waterways, trade, and settlement.

_____

_____

_____

_____

_____

Name _____   Class _____   Date _____

Follow the instructions below to practice test-taking on what you've learned from this section.

Multiple Choice Circle the best answer for each question from the options available.

1. Why is Europe called a "peninsula of peninsulas"?

   A Europe has water on all sides.

   B Europe is a peninsula and it contains peninsulas.

   C Europe has many waterways.

   D Europe has more than 24,000 miles of coastline.

2. What is a mistral?

   A an ocean current of warm water

   B land that has been taken from the seabed

   C water surrounded on three sides by land

   D a cold wind that brings cold, dry weather

3. About how long is the Mediterranean Sea?

   A 10,000 miles

   B 2,500 miles

   C 1,000 miles

   D 500 miles

4. What is one way explorers helped Europe's rulers in the 1400s?

   A They helped rulers improve fishing industries.

   B They helped rulers build empires and spread religious beliefs.

   C They built many inland waterways.

   D They helped rulers become allies.

5. What are fjords?

   A the German name for ports

   B very narrow bays found in Norway

   C polders built in the Netherlands

   D gaint walls used to hold back seawater

6. The Alps stretch from Austria and Italy to

   A Switzerland, Germany, and France.

   B France, Spain, and Portugal.

   C Hungary, Romania, and Ukraine.

   D Switzerland, Germany, and Poland.

7. What two major rivers flow through Europe?

   A Pyrenees and Appenines

   B Danube and Gdansk

   C Rhine and Danube

   D Rhine and Ukraine

8. What landform stretches across France, Belgium, Germany, and Poland to Russia?

   A the Western Uplands

   B the Northern European Plain

   C the Central Uplands

   D the Alpine region

9. What human activities have had an impact on the Mediterranean Sea's ecosystem?

   A building canals and waterways

   B overfishing and overdevelopment

   C exploring other lands

   D reclaiming seabeds

10. Centuries of misuse has caused the Mediterranean to lose

   A tourist income from its famous beaches.

   B a number of ports due to erosion.

   C large fish and red coral.

   D species of sea stars and sea urchins.

Document-Based Questions

The following excerpt is from the article **Nations Wrestle Over Ban on Tuna Trade** by David Braun, which appeared in *NATGEO News Watch* on March 5, 2010. Read the excerpt and answer the questions that follow.

The scarcer bluefin tuna becomes, the more its price rises, the more it is in demand as a delicacy. National Geographic Explorer-in-Residence and marine biologist Sylvia Earle once told me, "When the world is down to the last tuna, someone will be willing to pay a million dollars to eat it."

Constructed Response Read each question carefully and write your answer in the space provided.

11. According to the author, what is happening as a result of bluefin tuna becoming scarce?

_____

12. How does the expert, Sylvia Earle, think people will respond to the scarcity of bluefin tuna?

_____

Extended Response Read each question carefully and write your answer in the space provided.

13. Using the information you learned in Section 1.4, why do you think bluefin tuna is becoming scarce and how can we help to bring back the species?

_____

_____

PROJECTED POPULATION OF MEDITERRANEAN CITIES (IN MILLIONS)		
CITY	**1960**	**2015**
Athens, Greece	2.20	3.10
Barcelona, Spain	1.90	2.73
Istanbul, Turkey	1.74	11.72
Marseille, France	0.80	1.36
Rome, Italy	2.33	2.65

Source: UN, 2002

14. According to the chart, what trend does the information support?

_____

_____

_____

15. From what you have learned about human activities and ecosystems and given the information in the chart, what steps will people have to take to preserve the environment?

_____

_____

_____

© NGSP & HB

Name_____ Class_____ Date_____

Reading and Note-Taking **Track Details**

Use the Tree Diagram below to record details about ancient Greece's ideas about government as you read Section 2.1. Add branches for any additional details you may want to record.

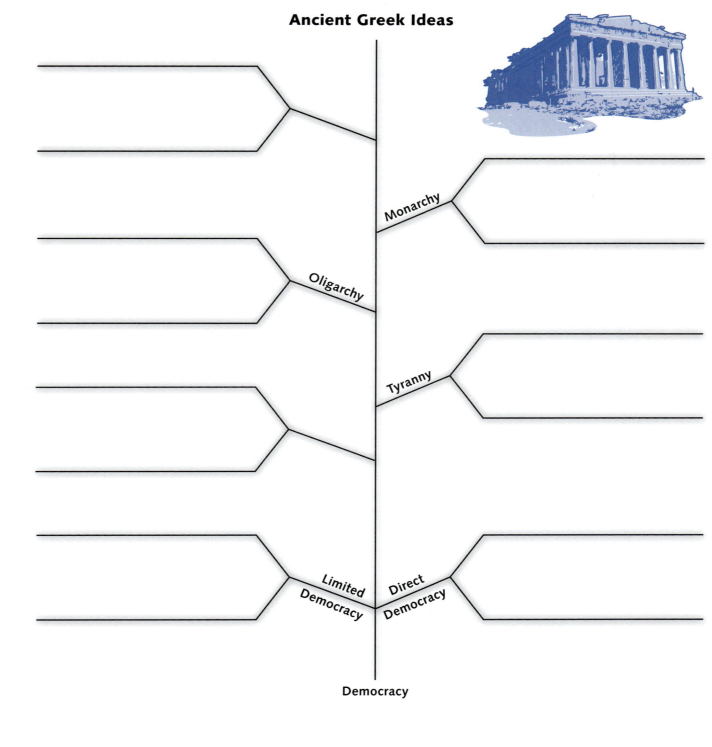

Ancient Greek Ideas

Monarchy

Oligarchy

Tyranny

Limited Democracy

Direct Democracy

Democracy

© NGSP & HB

SECTION **2** EARLY HISTORY

2.1 **Roots of Democracy**

Use with Europe Geography & History,
Section 2.1, *in your textbook.*

Vocabulary Practice

KEY VOCABULARY

- **city-state** n., a state that has its own government and consists of a city and the area around it
- **democracy** (deh-MAH-krah-see) n., a form of government in which people can influence law and vote for representatives

ACADEMIC VOCABULARY

- **aristocrat** (uh-RIS-toh-krat) n., a person who belongs to the highest social class

Write a sentence using the Academic Vocabulary word *aristocrat*.

_____

_____

Definition and Details Complete a Definition and Details Chart for the Key Vocabulary words *city-state* and *democracy*.

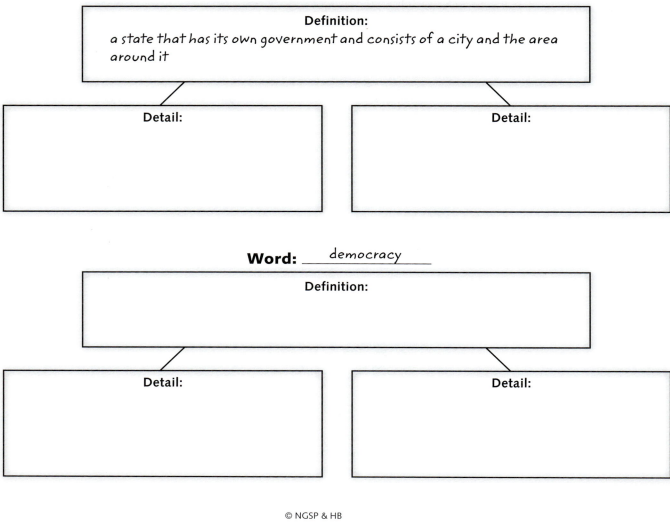

Word: _city-state_

Definition:
a state that has its own government and consists of a city and the area around it

Detail:

Detail:

Word: _democracy_

Definition:

Detail:

Detail:

SECTION **2** EARLY HISTORY

GeoActivity

Use with Europe Geography & History, Section 2.1, in your textbook.

Go to Interactive Whiteboard GeoActivities at
myNGconnect.com *to complete this activity online.*

NATIONAL GEOGRAPHIC
School Publishing

2.1 ROOTS OF DEMOCRACY

Analyze Primary Sources: Democracy

Read the two excerpts below to learn more about the influence of Greek thought on modern democracies. Then answer the questions.

Document 1: Foundations of Athenian Democracy

Pericles was a general, a politician, and a great orator, or public speaker, in ancient Athens. His leadership brought the city-state to its "golden age." His reforms gave power to ordinary people, not just to the wealthy.

We enjoy a form of government which is not in rivalry with the institutions of our neighbors, nay, we ourselves are rather an example to many than imitators of others. By name, since the administration is not in the hands of few but of many, it is called a democracy. And it is true that before the law and in private cases all citizens are on an equality.

—from Pericles' Funeral Oration, recorded by Thucydides

Marble bust of Pericles

Document 2: Foundations of American Democracy

We hold these truths to be self-evident, that all men are created equal, that they are endowed by their Creator with certain unalienable Rights, that among these are Life, Liberty, and the pursuit of Happiness.

—from the Declaration of Independence

The "Declaration Committee"—
Thomas Jefferson, Roger Sherman, Benjamin Franklin, Robert R. Livingston, and John Adams

---

1. **Summarize** According to Pericles, what defines a democracy?

_____

_____

2. **Identify** According to the Declaration of Independence, what fact is true of "all men"?

_____

_____

3. **Compare and Contrast** How are the thoughts on democracy expressed by Pericles and the authors of the Declaration of Independence similar?

_____

_____

4. **Make Inferences** Many of the leaders of the 13 colonies knew Greek and were familiar with Greek ideas on democracy. What influence might Greek thinkers have had on the founders of the U.S. government?

_____

_____

SECTION **2** EARLY HISTORY

2.2 Classical Greece

Use with Europe Geography & History, Section 2.2, *in your textbook.*

Reading and Note-Taking Pose and Answer Questions

Use the chart below to help you pose and answer questions about the many achievements of Classical Greece that you read about in Section 2.2.

Achievement	Question	Answer
Democracy	What did Pericles do to strengthen democracy?	He paid citizens to hold public office because as long as officials were unpaid, only the wealthy could afford to serve.
Architecture		
Philosophy		
Science		

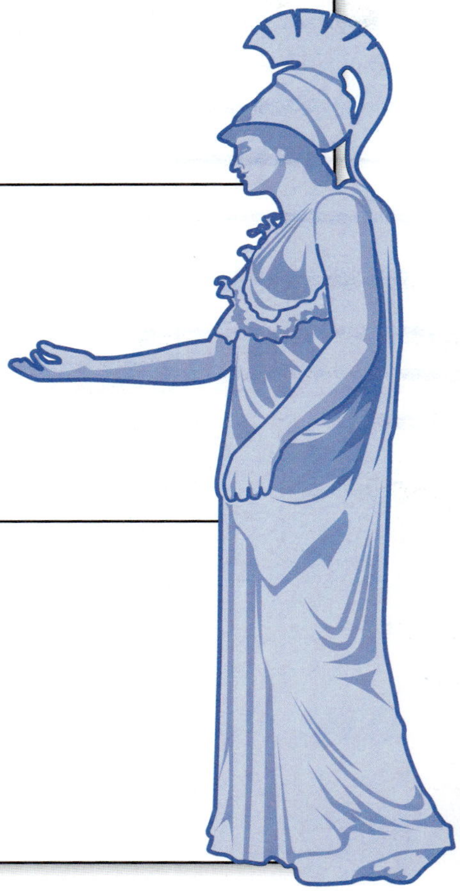

© NGSP & HB

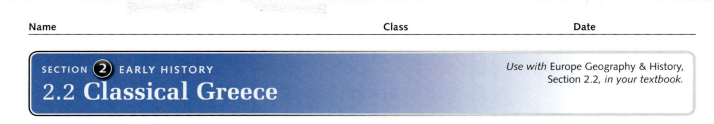

SECTION **2** EARLY HISTORY
2.2 Classical Greece

Use with Europe Geography & History, Section 2.2, *in your textbook.*

Vocabulary Practice

KEY VOCABULARY

• **golden age** n., a period of prosperity and cultural achievement

• **philosopher** (fil-AH-soh-fur) n., a person who closely examines basic questions about life and the universe to seek wisdom

Word Squares Complete Word Squares for the Key Vocabulary words *golden age* and *philosopher*.

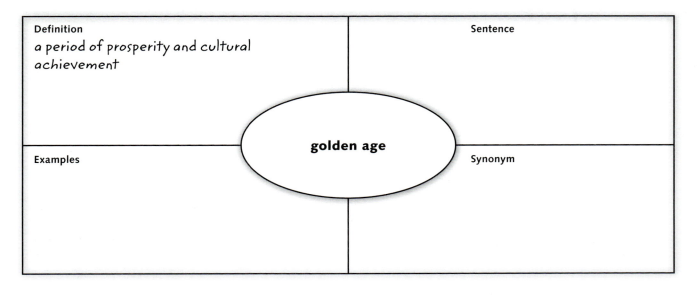

Definition
a period of prosperity and cultural achievement

Sentence

golden age

Examples

Synonym

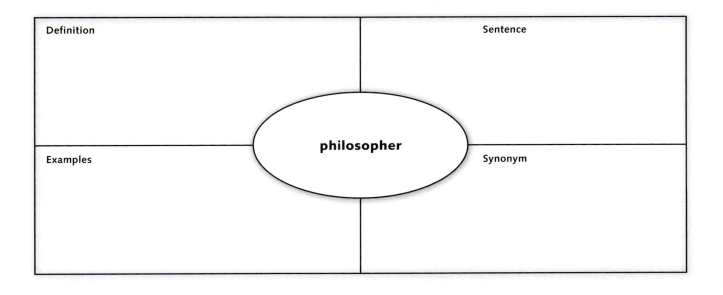

Definition

Sentence

philosopher

Examples

Synonym

Name _____ Class _____ Date _____

GeoActivity

2.2 CLASSICAL GREECE

Research Ancient Greek Contributions

Many aspects of culture and society today come from ideas and cultural traditions from classical Greece. In literature and drama, for example, the ancient Greeks wrote the first plays and developed the idea of comedy and tragedy. A Greek runner ran the first "marathon" to announce a military victory, while the idea of democracy was born in the Greek city-states. Greek ideas about art, architecture, health, science, and philosophy also influence modern thinking.

Follow these steps to research an ancient Greek contribution to modern society.

Step 1 Choose a Topic With your teammates, choose one of the following categories. To begin, write your topic on the lines at the center of the Idea Web at right.

art and architecture	literature and drama
government	science and philosophy
health	sports

Step 2 Share Your Knowledge What do the members of your team already know about your research topic? Take notes on a separate sheet of paper.

Step 3 Identify Research Sources Work with your teacher or a librarian to find sources of information about the topic. You can also use the research links at **Connect to NG.**

Step 4 Organize Your Research Organize the information your team has found. Then record the examples in the circles of the Idea Web.

Step 5 Share Your Findings Explain to the class what your team discovered about ancient Greek contributions.

Use with Europe Geography & History, Section 2.2, in your textbook.

Go to Interactive Whiteboard GeoActivities at
myNGconnect.com *to complete this activity online.*

NATIONAL GEOGRAPHIC
School Publishing

Idea Web

Contributions of Classical Greece in _____

SECTION **2** EARLY HISTORY

2.3 The Republic of Rome

Use with Europe Geography & History, Section 2.3, *in your textbook.*

Reading and Note-Taking **Find Main Idea and Details**

As you read Section 2.3, use a Main Idea and Details Diagram to take notes about the government of the Roman Republic. Include information from the text as well as from the Government of the Roman Republic chart.

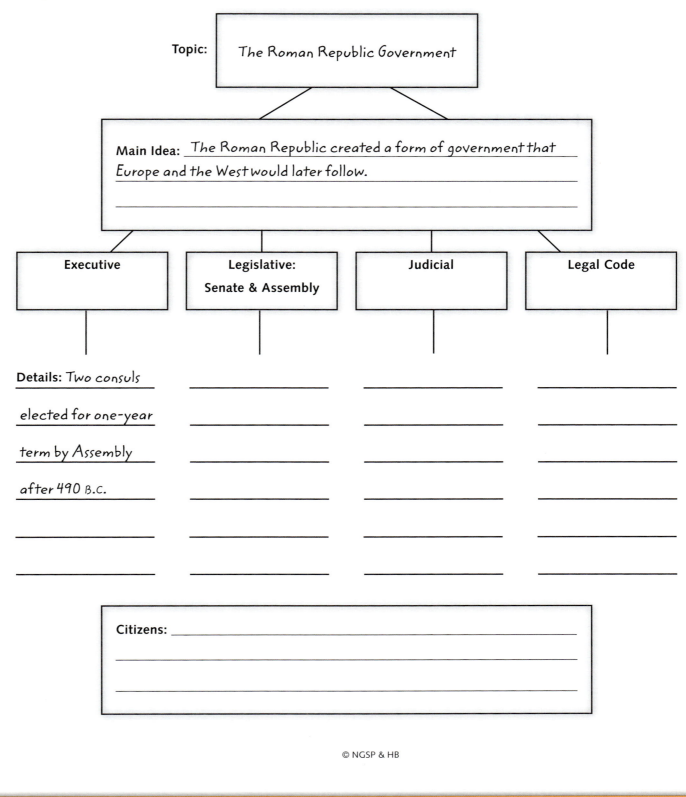

Topic: The Roman Republic Government

Main Idea: The Roman Republic created a form of government that Europe and the West would later follow. _____

| Executive | Legislative: Senate & Assembly | Judicial | Legal Code |

Details: Two consuls elected for one-year term by Assembly after 490 B.C. _____

Citizens: _____

SECTION **2** EARLY HISTORY

2.3 The Republic of Rome

Use with Europe Geography & History, Section 2.3, *in your textbook.*

Vocabulary Practice

KEY VOCABULARY

- **patrician** (pah-TRISH-uhn) n., a wealthy landowner and member of the highest social class
- **plebeian** (plih-BEE-yahn) n., a member of the common people of ancient Rome
- **republic** (rih-PUHB-lihk) n., a form of government in which people elect officials who govern according to law

ACADEMIC VOCABULARY

- **veto** (VEE-toh) v., to reject a proposed law officially

Write a sentence using the Academic Vocabulary word *veto.*

_____

_____

Compare/Contrast Paragraph Write a compare/contrast paragraph using the Key Vocabulary words *patrician, plebeian,* and *republic.* Explain what a republic is. Then compare and contrast the roles, duties, and rights of patricians and plebeians in the Roman Republic.

_____

_____

_____

_____

_____

_____

_____

_____

_____

_____

_____

_____

_____

SECTION **2** EARLY HISTORY

GeoActivity

2.3 THE REPUBLIC OF ROME

Compare Greek and Roman Governments

Learn more about Greek and Roman governments. Then complete a chart that compares and contrasts the two types of government.

Greek Democracy

Ancient Greece was divided into many small areas called city-states. Because Greek city-states were not bound together by any central ruler or empire, various forms of government emerged. One form was direct democracy. Athens, one of Greece's city-states, was governed by direct democracy during the fifth and fourth centuries B.C. Every person who was eligible to vote had a say in government decisions. Decisions supported by the majority became law.

The Roman Republic

Rome was originally a city-state. The powers of the Roman Senate, composed primarily of wealthy people, were balanced by the power of the Roman Assembly. The Assembly was a gathering of poorer Romans who met in the Forum, or hall. They could have a say in the election of magistrates and the enactment of new laws. As the Roman Republic expanded, it continued to have a central government. However, citizens who lived far from the city could not participate. The government decided that these citizens could elect representatives to attend the Assembly. In a republic, people do not vote directly for or against laws, but instead elect representatives to vote in their place.

1. **Create Charts** Use the information in the two passages to compare and contrast a direct democracy and a republic.

Use with Europe Geography & History, Section 2.3, in your textbook.

Go to Interactive Whiteboard GeoActivities at **myNGconnect.com** to complete this activity online.

NATIONAL GEOGRAPHIC
School Publishing

Direct Democracy and Republic

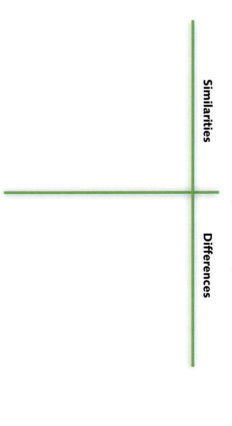

Similarities	Differences

2. **Identify** How did the Assembly ensure a more democratic government for the Roman Republic?

_____

_____

3. **Evaluate** Why do you think the United States government was formed as a republic, rather than as a direct democracy?

_____

_____

4. **Synthesize** What is a potential problem of a direct democracy, or "majority rule"? What is a potential problem of a republic?

_____

_____

SECTION **2** EARLY HISTORY

Use with Europe Geography & History, Section 2.4, *in your textbook.*

2.4 The Roman Empire

Reading and Note-Taking Sequence Events

As you read about the Roman Empire in Section 2.4, use a Sequence Chart to keep track of events. Write down specific dates, people, and places that contributed to the creation, rise, and decline of the Empire.

Creation of the Empire

46 B.C.: The Roman Senate makes Julius Caesar the first Roman Emperor

↓

Rise of the Empire

↓

Decline of the Empire

Describe the Roman Empire's legacy and its impact on modern life.

_____

_____

_____

_____

_____

SECTION **2** EARLY HISTORY
2.4 The Roman Empire

Use with Europe Geography & History,
Section 2.4, *in your textbook.*

Vocabulary Practice

KEY VOCABULARY
- **aqueduct** (AHK-wah-dukt) n., a structure, like a bridge or channel, that carries water across long distances over land
- **barbarian** (bar-BAYR-ih-yuhn) n., a member of an ancient German tribe

W-D-S Triangles Complete Word-Definition-Sentence Triangles for each Key Vocabulary word. Write the definition next to "D." Write a sentence using the word next to "S."

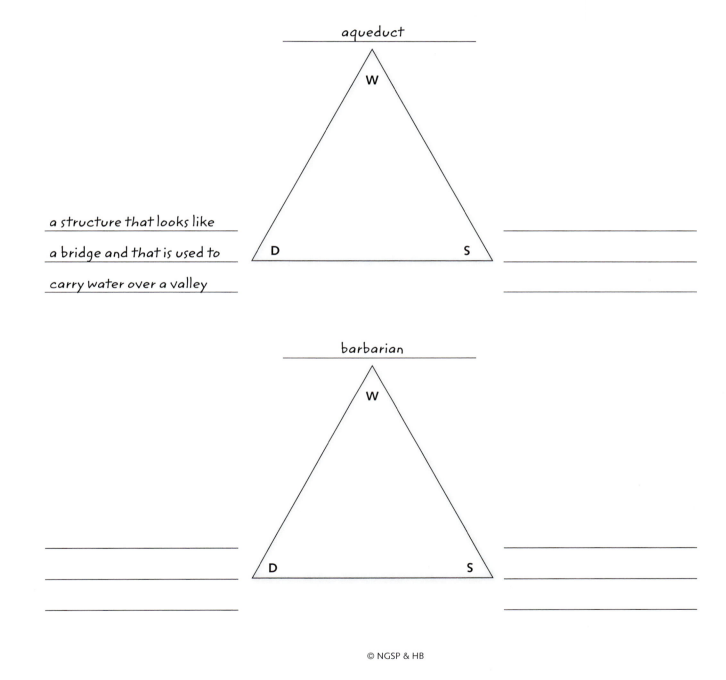

aqueduct

W

a structure that looks like

a bridge and that is used to

carry water over a valley

D   S

barbarian

W

D   S

Name _____ Class _____ Date _____

SECTION **2** EARLY HISTORY

GeoActivity

Use with Europe Geography & History, Section 2.4, in your textbook.

Go to Interactive Whiteboard GeoActivities at
myNGconnect.com *to complete this activity online.*

NATIONAL GEOGRAPHIC School Publishing

2.4 THE ROMAN EMPIRE

Analyze the Roots of Modern Languages

Latin was the language of the ancient Romans. It is also the basis of many languages, including modern Spanish, French, and Italian. Other languages, including English, are heavily influenced by Latin. The following chart contains Latin roots and prefixes that are found in many English words. Study the chart and then answer the questions.

LATIN ROOT	MEANING	EXAMPLE
spec, spect	to look at	**spectator** n., a person who watches
vers, vert	to turn	**reverse** v., to go backward
bon	good	**bonus** adj., extra
leg	law	**legal** adj., allowed by law
ped	foot	**pedestrian** n., a person who is walking
dom	house	**domicile** n., the place where a person lives
dis-	not, apart	**disapprove** v., to believe something is wrong
sub-	under	**submerge** v., to go under

1. Make Connections From the Latin roots in the table, choose the one related to each of these words. (Use a dictionary to check their meaning.)

a. distrust _____

b. bon voyage _____

c. biped _____

d. pedal _____

e. subterranean _____

f. inspect _____

2. Draw Conclusions

a. From its Latin root, what do you think is the function of a *legislature*?

_____

b. The Latin scientific name for a pet cat is *Felis domesticus*. What Latin root is in this name? What do you think the name means?

_____

c. What do you think a *pedometer* measures?

_____

d. The Latin word for "sea" is *mare*. Where does a *submarine* travel?

_____

e. If you are averse to an idea, are you for or against it?

_____

3. Synthesize How many English words can you create from the roots and prefixes in the table? Check your answers with a dictionary.

_____

© NGSP & HB

Name ____________________ Class ____________ Date ____________

SECTION 2 EARLY HISTORY

2.5 Middle Ages and Christianity

Use with Europe Geography & History, Section 2.5, *in your textbook.*

Reading and Note-Taking Summarize Information

Use the chart below as you read Section 2.5 to take notes about the influences on Western Europe during the Middle Ages. Then write a summary of the information in your chart.

Subject
The Middle Ages

The Roman Catholic Church	The Feudal System	The Growth of Towns
Helped unite people during the		
Middle Ages		
Played a leading role in		
government by collecting taxes		

Summary

SECTION **2** EARLY HISTORY

2.5 Middle Ages and Christianity

Use with Europe Geography & History,
Section 2.5, *in your textbook.*

Vocabulary Practice

KEY VOCABULARY

- **feudal** (FYOO-duhl) **system** n., a social system that existed in Europe during the Middle Ages in which people worked for nobles who in return gave them protection and the use of land

- **serf** n., a person who farmed the land owned by a noble in return for shelter and protection

Vocabulary Pyramid Use a Vocabulary Pyramid to show how the feudal system worked in the Middle Ages. Define the Key Vocabulary word *feudal system* in your own words. Then write which social group was at each level in the pyramid and the role that the group had. Use the Key Vocabulary word *serf* in your pyramid.

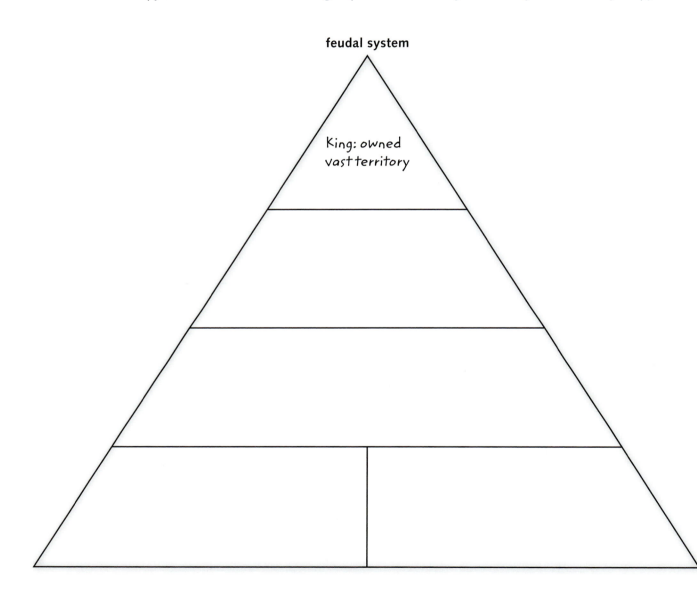

feudal system

King: owned vast territory

Name _____ Class _____ Date _____

GeoActivity

Categorize Effects of the Crusades

The Crusades caused great changes in European culture and history. Read the passage below about the effects of the Crusades. Then answer the questions.

Between 1096 and 1291, thousands of Europeans—kings, nobles, peasants, and pilgrims—traveled to the eastern Mediterranean as Crusaders. There were four major Crusades, or military expeditions, and several lesser ones. The goal of each was to free Christian sites, such as Jerusalem (in present-day Israel), from Muslim rule. Many Crusaders died in battle or from disease or hunger on these long journeys.

Cultural Diffusion and the Crusades

Europe in the Middle Ages was far behind the Islamic world in learning, science, and technology. Most Europeans, even nobles, had never traveled far from home. Now people from small villages saw great cities with beautiful marble buildings. The Crusaders tasted new foods and encountered new ideas. Those who survived never forgot what they experienced. For example, some returning Crusaders built massive stone castles, like those they had seen in the East.

Trade in the Mediterranean prospered during the Crusades. Arab technology improved shipbuilding. Luxury goods such as sugar, silk, flax, spices, and shoes were sent to Europe from the East. The Italian trading cities Venice and Genoa grew rich and powerful. Merchants, part of a growing middle class, became wealthy.

The Crusades also changed European society by weakening the feudal system. Many nobles never returned home from the Crusades. Others spent their fortunes on the journey. Their lands became the property of the king, centralizing his power. The power and wealth of the Pope and the Roman Catholic Church also grew during this period.

Use with Europe Geography & History, Section 2.5, in your textbook.

Go to Interactive Whiteboard GeoActivities at **myNGconnect.com** to complete this activity online.

☐ **NATIONAL GEOGRAPHIC**
School Publishing

1. Categorize Analyze the information in the passage to categorize different effects the Crusades had on European history and culture. Then fill in the chart.

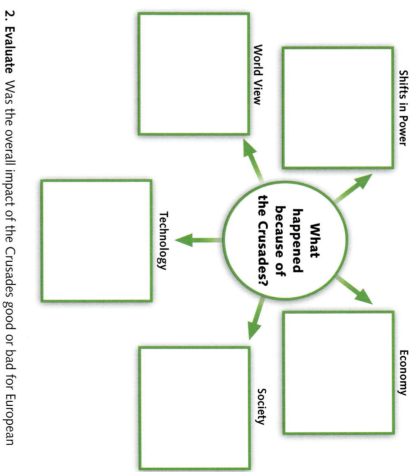

Shifts in Power

World View

What happened because of the Crusades?

Technology

Economy

Society

2. Evaluate Was the overall impact of the Crusades good or bad for European culture? Explain your answer.

_____

© NGSP & HB

SECTION **2** EARLY HISTORY
2.6 Renaissance and Reformation

Use with Europe Geography & History, Section 2.6, *in your textbook.*

Reading and Note-Taking Analyze Cause and Effect

As you read Section 2.6, note the factors that led to the Renaissance. Then use the Cause and Effect Chart to show the effects that the Renaissance had on Europe.

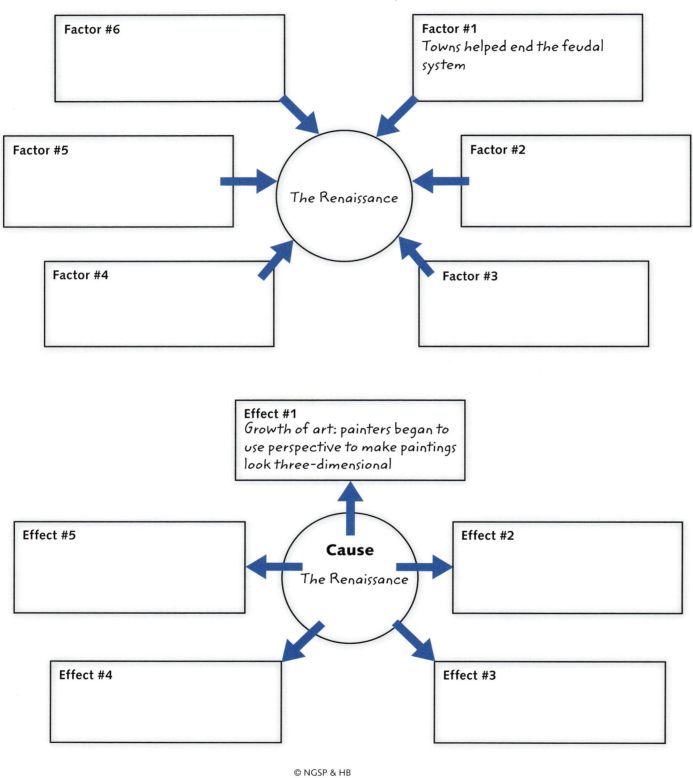

Factor #6

Factor #1
Towns helped end the feudal system

Factor #5

The Renaissance

Factor #2

Factor #4

Factor #3

Effect #1
Growth of art: painters began to use perspective to make paintings look three-dimensional

Effect #5

Cause
The Renaissance

Effect #2

Effect #4

Effect #3

© NGSP & HB

Name _____ Class _____ Date _____

Vocabulary Practice

KEY VOCABULARY
- **indulgence** (ihn-DULL-jenz) n., a payment for forgiveness of sin, sold by the Church
- **perspective** (puhr-SPEK-tihv) n., a technique that makes a painting look as if it has three dimensions

Definition Chart Complete a Definition Chart for the Key Vocabulary words *indulgence* and *perspective*.

Word	indulgence	perspective
Definition		
In Your Own Words		
Write a Sentence		

© NGSP & HB

Name _____ Class _____ Date _____

GeoActivity

2.6 RENAISSANCE AND REFORMATION

Use with Europe Geography & History, Section 2.6, in your textbook.

Go to Interactive Whiteboard GeoActivities at
myNGconnect.com to complete this activity online.

NATIONAL GEOGRAPHIC
School Publishing

Map the Protestant Reformation

Martin Luther was not the only European who criticized the Roman Catholic Church. People in other parts of Europe also wanted to change or reform some church practices. However, opposing the powerful Church could be dangerous. In some places, authorities persecuted heretics (HAIR-uh-tix), or people who disagreed with the teachings of the Church. In other places, however, rulers encouraged the new movement. For example, some English monarchs actively supported a Protestant state church. In other parts of Europe, wars broke out between people with differing beliefs.

Use information in the chart to fill in the map. Mark where each Protestant movement started. Then shade in the areas where it spread. Use a different pattern or color for each group and show it in a map legend.

RELIGION	FOUNDER/ LEADER	ORIGINAL LOCATION	SPREAD TO
Anglicanism (Church of England)	King Henry VIII	London, England	Throughout England
Calvinism	John Calvin	Geneva, Switzerland	France, Scotland, the Netherlands
Lutheranism	Martin Luther	Wittenberg, Germany	Holy Roman Empire, Sweden, Denmark, Norway

1. Interpret Maps Which early Protestant movement spread to the largest area in Europe? Which was limited to one country?

_____

_____

_____

MAJOR PROTESTANT RELIGIONS, c. 1650

Map legend:
Major Protestant Religions, c. 1650
☐ Anglicanism
☐ Calvinism
☐ Lutheranism

© NGSP & HB

2. Make Inferences Protestant movements had little influence in southern Europe. Why might this have been the case?

_____

_____

_____

Name _____ Class _____ Date _____

Follow the instructions below to review what you have learned in this section.

Vocabulary Next to each vocabulary word, write the letter of the correct definition.

1._____ democracy

2._____ golden age

3._____ philosopher

4._____ patrician

5._____ plebeian

6._____ aqueduct

7._____ indulgence

A. a person who closely examines basic questions to seek wisdom

B. a wealthy landowner and member of the highest social class

C. a structure that carries water across long distances over land

D. a form of government in which people can influence law and vote for representatives

E. a member of the common people of ancient Rome

F. a payment for forgiveness of sin, sold by the Church

G. a period of prosperity and cultural achievement

Main Ideas Use what you've learned about the early history of Europe to answer these questions.

8. **Identify** What were the two largest and most important Greek city-states?

_____

_____

9. **Identify and Explain** What goals did Pericles have for Greece and how did he achieve his goals?

_____

_____

10. **Summarize** How did Greek culture spread?

_____

_____

11. **Compare and Contrast** How were the executive branch and the judicial branch of the Roman Republic similar and how were they different?

_____

_____

12. **Explain** In your own words, what does *The Roman Way* mean?

_____

_____

13. **Describe** What marked the beginning of the fall of the Roman Empire?

_____

_____

14. **Analyze Cause and Effect** What happened as towns in Western Europe grew and trade and other businesses developed?

_____

_____

15. **Connect** How did the invention of the printing press influence the Renaissance?

_____

_____

© NGSP & HB

Focus Skill: Draw Conclusions

Answer the questions below to draw conclusions about how European thought shaped Western civilization.

16. How did the location of Greek city-states contribute to each having its own community and government?

_____

_____

17. Why do you think democracy thrived in Athens but not in Sparta?

_____

_____

18. What role do you think Socrates and Plato played in the continung influence ancient Greece has today?

_____

_____

19. What can you conclude about the relationship between Roman patricians and plebeians before 490 B.C.?

_____

_____

20. Why do you think Julius Caesar's rule in Rome was so significant?

_____

_____

21. How do you think the Roman Empire remained so powerful for so long?

_____

_____

22. What difference did it make during the Renaissance for writers to write in the vernacular, or the common language spoken in a region?

_____

_____

23. Why do you think the Renaissance was a time of "rebirth"?

_____

_____

Synthesize: Answer the Essential Question

How did European thought shape Western civilization? Consider information you have learned about forms of government, communities, rulers, and artistic expression in the early history of Europe.

_____

_____

_____

_____

_____

SECTION **2** EARLY HISTORY

2.1–2.6 Standardized Test Practice

Use with Europe Geography & History, Sections 2.1–2.6, *in your textbook.*

Follow the instructions below to practice test-taking on what you've learned from this section.

Multiple Choice Circle the best answer for each question from the options available.

1. What was possibly the most lasting achievement of Greek civilization?

 A laying the groundwork for democracy
 B always having great leaders
 C enabling transportation and communication
 D having many city-states

2. Alexander the Great spread ideas about what?

 A Macedonian approaches to government, religion, and power
 B his father's accomplishments in math, music, and history
 C ways to rule and and extend an empire
 D Greek ideas about democracy, science, and philosophy

3. In reality, who founded Rome?

 A the Romans
 B the people of Latium, or Latins
 C the Etruscans
 D the plebeians

4. What was significant about the Twelve Tables of 450 B.C.?

 A It was a plan for rebelling against Etruscan King Tarquin.
 B Laws enacted gave Roman women the right to vote.
 C They spelled out the responsibilities of Roman citizens.
 D They marked the beginning of the Roman Empire.

5. How did Julius Caesar help the poor?

 A He created farmland for them.
 B He gave money to them.
 C He built shelters for them.
 D He gave voting rights to them.

6. What is one of Rome's great engineering legacies?

 A the minting of Roman coins
 B use of marble for buildings and statues
 C the network of roads connecting the empire
 D advances in weaponry against invaders

7. What period lasted from 500 to 1500?

 A the Crusades
 B the Renaissance
 C the Middle Ages
 D the Reformation

8. What system provided security for kingdoms after Charlemagne?

 A the Franks system
 B the Roman Catholic Church
 C the system of towns
 D the feudal system

9. How did Gutenberg's invention empower Europeans?

 A It gave them a new invention to trade.
 B It gave them access to knowledge.
 C It gave them new techniques for art and architecture.
 D It gave them stronger armies.

10. What is perspective?

 A a painting technique
 B a farming method
 C a battle strategy
 D a religious meditation

Document-Based Questions

The following excerpt is taken from an article **The Hanseatic League: Europe's First Common Market** by Edward Von Der Porten in *National Geographic* in the October 1994 issue. Read the excerpt and answer the questions that follow.

> *In the countryside millions of landless serfs were bound to drudgery and servitude* [hard work and little freedom]. *The urban Hansards* [members of a trade group], *by contrast . . . guarded their independence. "Town air is freedom" was their civic maxim* [saying]. *If a serf could escape the countryside and survive a year and a day within a town's walls, he could no longer be claimed as property by a nobleman.*

Constructed Response Read each question carefully and write your answer in the space provided.

11. According to the article, what was life like for a landless serf?

_____

12. What happened when a serf survived a year and a day within a town's walls?

_____

Extended Response Read each question carefully and write your answer in the space provided.

13. Using what you learned from Section 2.5 and this excerpt, what can you conclude about serfs who had "escaped"?

_____

_____

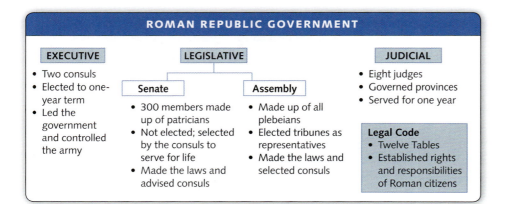

ROMAN REPUBLIC GOVERNMENT

EXECUTIVE
- Two consuls
- Elected to one-year term
- Led the government and controlled the army

LEGISLATIVE

Senate
- 300 members made up of patricians
- Not elected; selected by the consuls to serve for life
- Made the laws and advised consuls

Assembly
- Made up of all plebeians
- Elected tribunes as representatives
- Made the laws and selected consuls

JUDICIAL
- Eight judges
- Governed provinces
- Served for one year

Legal Code
- Twelve Tables
- Established rights and responsibilities of Roman citizens

14. What are the three branches of the Roman Republic government?

_____

15. How did the Roman Republic government address the roles of the classes?

_____

_____

© NGSP & HB

SECTION **3** EMERGING EUROPE
3.1 Exploration and Colonization

Use with Europe Geography & History, Section 3.1, *in your textbook.*

Reading and Note-Taking Outline and Take Notes

As you read about European exploration and colonization in Section 3.1, use an outline to help you take notes.

Exploration and Colonization

I. *Purpose of Exploration* _____

   A. *to find gold and establish trade with Asia* _____

   B. _____

II. _____

   A. _____

   B. _____

III. _____

   A. _____

   B. _____

   C. _____

   D. _____

IV. _____

   A. _____

   B. _____

   C. _____

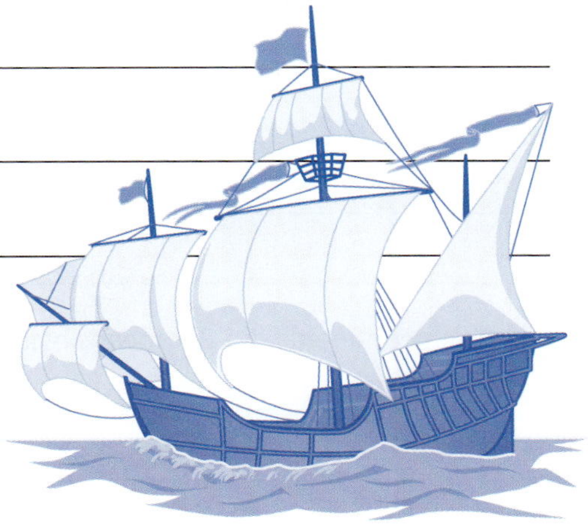

© NGSP & HB

SECTION ❸ EMERGING EUROPE

3.1 Exploration and Colonization

Use with Europe Geography & History, Section 3.1, *in your textbook.*

Vocabulary Practice

KEY VOCABULARY

- **colony** n., an area settled and controlled by a distant country
- **navigation** n., the skill of steering modes of transportation and reading maps to get to particular locations

ACADEMIC VOCABULARY

- **convert** (kuhn-VERT) v., to change from one religion to another

Write a sentence using any form of the Academic Vocabulary word *convert*.

_____

_____

Words in Context Follow the directions below for using the Key Vocabulary in context.

1. Use a form of the word *colony* in a sentence about North America.

_____

_____

_____

2. Write a sentence defining the word *navigation*.

_____

_____

_____

3. Write a sentence about *colonies* in South America.

_____

_____

_____

4. Write a sentence describing Prince Henry's school of *navigation* and the effect it had on European exploration.

_____

_____

_____

5. On which continents did Europeans establish colonies by 1650?

_____

_____

_____

☐ NATIONAL GEOGRAPHIC School Publishing

SECTION ③ EMERGING EUROPE

GeoActivity

3.1 EXPLORATION AND COLONIZATION

Compare European Explorers

Read the descriptions below to learn more about European explorers. Then complete the chart to summarize information about each one.

Use with Europe Geography & History, Section 3.1, in your textbook.

Go to Interactive Whiteboard GeoActivities at **myNGconnect.com** *to complete this activity online.*

John Cabot (1450–1499), an Italian navigator and explorer, was sponsored by King Henry VII of England to search for new lands. Cabot's voyages in 1497 and in 1498 helped establish British claims in Canada.

Vasco Núñez de Balboa (1475–1519), a Spanish conquistador, is known as the first European to see the eastern shore of the Pacific Ocean in 1513. Balboa claimed the Pacific Ocean and its shores for Spain.

Pedro Álvares Cabral (1467/68–1520), a Portuguese navigator and explorer, is generally credited with discovering Brazil for Portugal. In 1500, King Manuel of Portugal selected Cabral to follow the route taken earlier by Vasco da Gama and to further his conquest.

Amerigo Vespucci (1454–1512), an Italian explorer, navigator, and mapmaker, is the explorer after which the Americas were named. Vespucci took part in at least two voyages to South America, one for Spain (1499) and the other for Portugal (1501).

Jacques Cartier (1491–1557), a French navigator and explorer, explored the Canadian coast and St. Lawrence River between 1534 and 1542. His travels there served as the basis for French claims in North America.

1. Create Charts Use the information you learned about explorers to complete the chart.

EXPLORER	COUNTRY OF ORIGIN	COUNTRY FUNDING EXPEDITION	RESULTS/ ACHIEVEMENTS
Cabot			
Balboa			
Cabral			
Vespucci			
Cartier			

2. Summarize England, Spain, France, and Portugal sponsored the trips of these explorers. What was it that these countries wanted to gain as a result of the explorations?

_____

3. Make Inferences What personal characteristics and abilities would an explorer need to have during this period?

_____

4. Synthesize What did all of the expeditions have in common?

_____

© NGSP & HB

SECTION **3** EMERGING EUROPE

Use with Europe Geography & History, Section 3.2, *in your textbook.*

3.2 **The Industrial Revolution**

Reading and Note-Taking Form and Support Opinions

As you read Section 3.2, use the chart below to help you form and support your opinions about the Industrial Revolution. Read the question below, and then list the pros and cons of the Industrial Revolution. Finally, write your opinion and use evidence from the text and your notes to support it.

Question:

In your opinion, were the changes of the Industrial Revolution positive, negative, or both?

Pros	Cons
Standards of living rose.	Factory workers often faced harsh conditions.

Opinion

© NGSP & HB

Name _____ Class _____ Date _____

Vocabulary Practice

KEY VOCABULARY

- **factory system** n., a system in which each person works on a small part of a larger project
- **textile** (TEHKS-tyl) n., cloth

Cause and Effect Paragraph Write a paragraph explaining the impact that the Industrial Revolution had on the textile industry. Use the Key Vocabulary words *factory system* and *textile* in your paragraph and define what they mean. Describe the way textiles were made before and after the Industrial Revolution.

_____

_____

_____

_____

_____

_____

_____

_____

_____

_____

_____

_____

_____

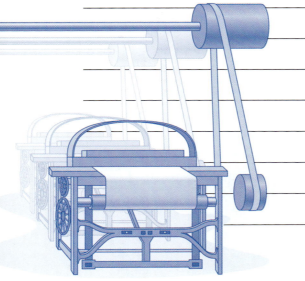

_____

_____

© NGSP & HB

Name _____ Class _____ Date _____

GeoActivity

3.2 THE INDUSTRIAL REVOLUTION

Evaluate Industrial Revolution Inventions

Read the descriptions below to learn about inventions of the Industrial Revolution. As a class or in small groups, debate which invention was the most important. Use the chart to organize your thoughts and support your arguments.

Reliable Steam Engine In 1775, the Scottish inventor James Watt invented the first reliable steam engine. He improved the efficiency of the Newcomen steam engine, which used a lot of coal and lost steam. Now a faster pace of transportation and faster machines were possible. Because of Watt, steam became the energy source that would power the Industrial Revolution.

Cotton Gin In 1797, Eli Whitney invented the cotton gin. He was an inventor and mechanical engineer from the United States. The cotton gin sped up cotton production. It could clean seeds from cotton fibers faster than people could clean them by hand. This machine helped the United States meet the increased demand for cotton after English factories started spinning thread mechanically.

Telephone Alexander Graham Bell, a Scottish-born American scientist and audiologist, had an idea. It was to create a machine that would transmit sound using electricity and wires. In 1876 Bell invented the telephone.

Use with Europe Geography & History, Section 3.2, in your textbook.

Go to Interactive Whiteboard GeoActivities at **myNGconnect.com** to complete this activity online.

Assembly Line Following the success of his Model T, the U.S. industrialist Henry Ford faced a problem: how to produce a large volume of motor vehicles at a low price. In 1913, his solution was to create the moving assembly line. The assembly line consisted of a conveyer system and workers who performed repetitive tasks. The plan worked. By 1926, a car that had cost $950 in 1909 sold for $290.

1. **Create Charts** Use what you learned about inventions of the Industrial Revolution to complete the chart.

INVENTION	HOW IT WAS USED	IMPACT ON SOCIETY
Reliable steam engine		
Cotton gin		
Telephone		
Assembly line		

2. **Turn and Talk** With a partner, talk about the effects these inventions had on society—on work life, on home life, and on businesses. For each invention, think beyond its first use. For example, what kind of impact would the assembly line have had on the factory system in general, not solely for automobile factories?

3. **Debate** Which invention do you think was the most important? Support your argument.

_____

_____

_____

© NGSP & HB

Name _____ Class _____ Date _____

SECTION **3** EMERGING EUROPE

3.3 The French Revolution

Use with Europe Geography & History,
Section 3.3, *in your textbook.*

Reading and Note-Taking Analyze Cause and Effect

As you read Section 3.3, use a Cause and Effect Map to help you analyze the causes and effects of the French Revolution. Write down the roots of the Revolution. Then write down the effects that each root helped to cause.

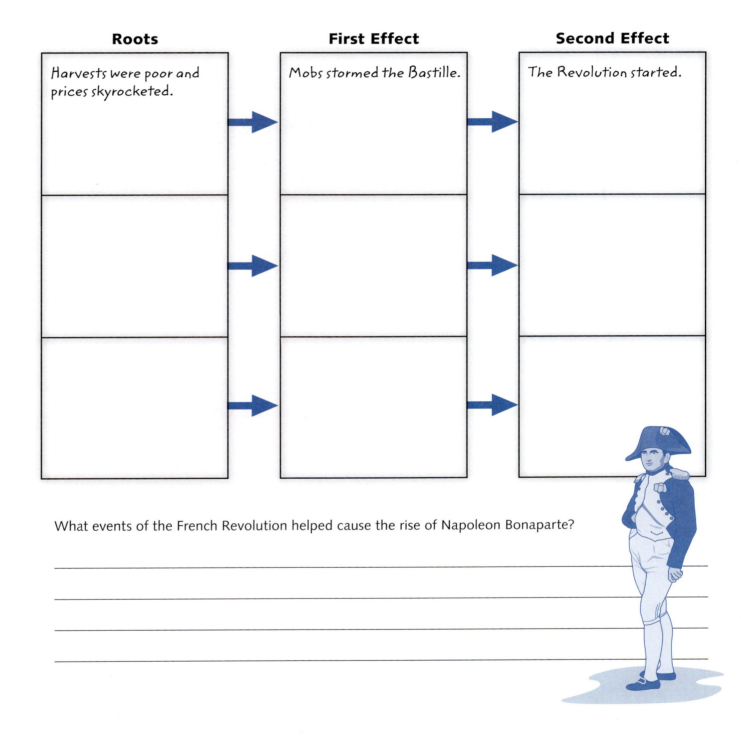

Roots	First Effect	Second Effect
Harvests were poor and prices skyrocketed.	Mobs stormed the Bastille.	The Revolution started.

What events of the French Revolution helped cause the rise of Napoleon Bonaparte?

_____

_____

_____

_____

© NGSP & HB

Name Class Date

SECTION 3 EMERGING EUROPE
3.3 The French Revolution

Use with Europe Geography & History, Section 3.3, *in your textbook.*

Vocabulary Practice

KEY VOCABULARY

- **guillotine** (GHEE-uh-teen) n., a machine for beheading as a form of execution
- **radical** n., a person who favors extreme changes in government

I Read, I Know, and So Complete the graphic organizers below for each Key Vocabulary word.

I Read
The Jacobins used a machine called the guillotine to cut off the heads of an estimated 40,000 people.

guillotine

I Know	And So

I Read

radical

I Know	And So

BLACK-AND-WHITE COPY READY

SECTION ③ EMERGING EUROPE

GeoActivity

Use with Europe Geography & History, Section 3.3, in your textbook.

Go to Interactive Whiteboard GeoActivities at
myNGconnect.com *to complete this activity online.*

☐ **NATIONAL GEOGRAPHIC**
School Publishing

3.3 THE FRENCH REVOLUTION

Map Napoleon's Empire

Napoleon Bonaparte was a brilliant young general who led the French army across Europe. Over a period of 15 years, he conquered some rulers and made treaties with others. Read the account of how Napoleon built an empire.

Napoleon Changes the Map of Europe

As part of his war against Austria, Napoleon moved his armies into northern Italy. A treaty in 1801 extended the borders of France to the Alps, the Rhine River, and the Pyrenees. That same year, an agreement with the Pope gave France control over the Papal States around Rome.

In December 1805, Napoleon won a great victory against Russia and Austria at Austerlitz and gained control over their territories in Italy and Germany. The powerful states of Austria and Prussia then gave up the fight, and Napoleon made himself head of the Kingdom of Italy.

In 1806, Napoleon made one of his relatives king of the Kingdom of Naples. He also pulled together several German states to create the Confederation of the Rhine. In 1807, a treaty with Czar Alexander of Russia created the Grand Duchy of Warsaw (Poland), which lay between Prussia and the Russian border. By 1810, Napoleon had redrawn the map of Europe. He (and his family members) ruled Spain, the Netherlands, Italy, and Switzerland.

1. **Create Maps** On the map at right, use arrows to show Napoleon's path while creating his empire. Then use one color to shade in the territories that came under his control. Label them in the legend.

2. **Interpret Maps** Find Austerlitz on the map. What does Napoleon's victory there indicate about his military skill?

_____

_____

NAPOLEON'S EMPIRE, 1810

☐ Napoleon's empire
☐ Napoleon's forces

0   200   400 Miles
0   200   400 Kilometers

ATLANTIC OCEAN

GREAT BRITAIN
North Sea
DENMARK
SWEDEN
Baltic Sea
NETHERLANDS
PRUSSIA
CONFEDERATION OF THE RHINE
GRAND DUCHY OF WARSAW
FRENCH EMPIRE
Rhine R.
• Austerlitz
AUSTRIAN EMPIRE
RUSSIAN EMPIRE
SWITZERLAND
KINGDOM OF ITALY
ILLYRIAN PROVINCES
Danube R.
SPAIN
CORSICA
PAPAL STATES
KINGDOM OF SARDINIA
MONTENEGRO
OTTOMAN EMPIRE
Mediterranean Sea
KINGDOM OF SICILY
KINGDOM OF NAPLES
Black Sea
Aegean Sea

N S E W

3. **Make Inferences** Great Britain was Napoleon's greatest enemy, but he never captured any British territory. How does geography explain this?

_____

_____

SECTION 3 EMERGING EUROPE
3.4 Declarations of Rights

Use with Europe Geography & History,
Section 3.4, *in your textbook.*

Reading and Note-Taking Analyze Primary Sources

Use a Main Idea Diagram to help you analyze primary sources as you read Section 3.4. Read the excerpt below. Find the main idea and supporting details and write them in your own words in the diagram. Then create and fill in a Main Idea Diagram for Document 2 in your textbook.

Document 1

Declaration of Independence (July 4, 1776)

> *We hold these truths to be self-evident, that all men are created equal, that they are endowed [provided] by the Creator with certain unalienable [guaranteed] Rights, that among these are Life, liberty, and the pursuit of Happiness; that, to secure these rights, Governments are instituted among Men, deriving their just powers from the consent of the governed.*

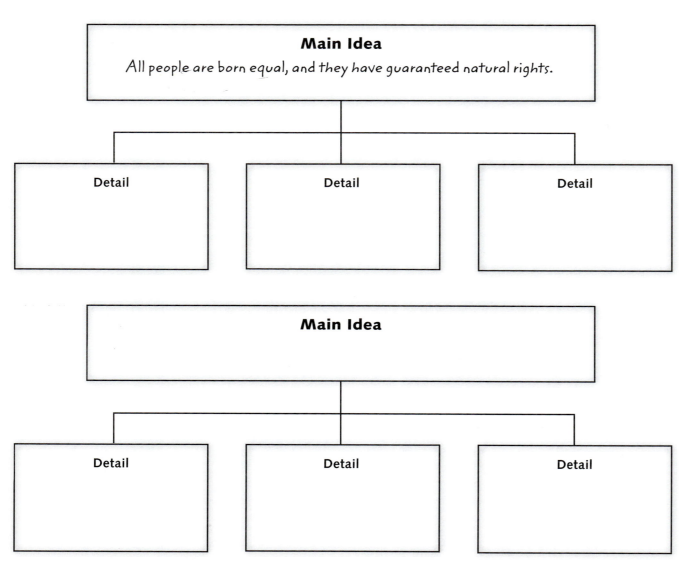

Main Idea

All people are born equal, and they have guaranteed natural rights.

| Detail | Detail | Detail |

Main Idea

| Detail | Detail | Detail |

© NGSP & HB

SECTION **3** EMERGING EUROPE
3.4 Declarations of Rights

Use with Europe Geography & History, Section 3.4, *in your textbook.*

Vocabulary Practice

KEY VOCABULARY

• **apartheid** (uh-PAHRT-hyt) n., a former social system in South Africa that denied black South Africans their rights

• **natural rights** n., rights that all people possess at birth

Meaning Map Complete a Meaning Map for each Key Vocabulary word.

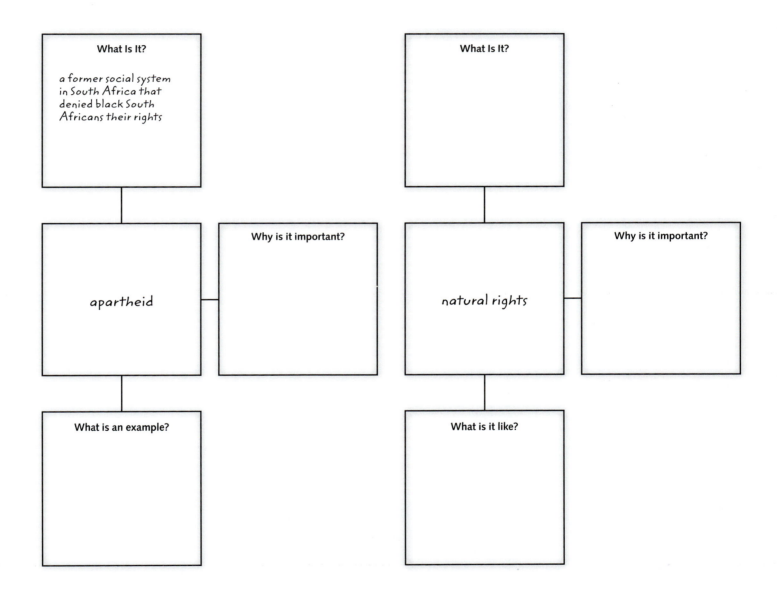

What Is It?

a former social system in South Africa that denied black South Africans their rights

apartheid

Why is it important?

What is an example?

What Is It?

natural rights

Why is it important?

What is it like?

Name _____ Class _____ Date _____

Use with Europe Geography & History, Section 3.4, in your textbook.

Go to Interactive Whiteboard GeoActivities at
myNGconnect.com to complete this activity online.

NATIONAL
GEOGRAPHIC
School Publishing

SECTION **3** EMERGING EUROPE

GeoActivity

3.4 DOCUMENT-BASED QUESTION: DECLARATIONS OF RIGHTS

Analyze Primary Sources: Women's Rights

During the Enlightenment, people claimed new rights and freedoms after centuries of strict rule by monarchs. Nearly all the leaders of those movements were men. However, women also spoke out to ensure that these rights could be enjoyed by both sexes. Analyze the following primary sources and answer the questions.

Document 1: Mary Wollstonecraft on the Rights of Women
Mary Wollstonecraft, a British woman, was inspired by the revolution in France to publish *A Vindication [Justification] of the Rights of Woman* in 1792. She criticized the ways in which all the new independence movements ignored the inequality that women faced.

If the abstract [natural] rights of man will bear discussion and explanation, those of woman, by a parity [similarity] of reasoning, will not shrink from the same test. . . .

Consider . . . whether, when men contend [struggle] for their freedom, and to be allowed to judge for themselves respecting their own happiness, it be not inconsistent and unjust to subjugate [control] women . . . ? Who made man the exclusive judge, if woman partake with him of the gift of reason?

—from the dedication of *A Vindication of the Rights of Woman*, 1792

1. **Explain** Why does Wollstonecraft believe women should have the same rights as men?

_____
_____
_____

Document 2: Abigail Adams on the Formation of a New Government
Abigail Adams wrote letters to her husband, John, while he was serving in the Continental Congress during the American Revolution. She was ahead of her time in her thoughts regarding women's rights.

I long to hear that you have declared an independency—and by the way, in the new Code of Laws which I suppose it will be necessary for you to make, I desire you would Remember the Ladies, and be more generous and favourable to them than your ancestors. . . . If particular care and attention is not paid to the Ladies we are determined to foment [form] a Rebellion, and will not hold ourselves bound by any Laws in which we have no voice, or Representation.

—from a letter to John Adams, March 31, 1776

2. **Interpret** What does Adams say will happen if women's rights are not considered in the laws of the new country? Why will this happen?

_____
_____

3. **Make Inferences** Did women have rights under the British national or colonial governments? How can you tell?

_____
_____

4. **Synthesize** Reread the the first two documents in your textbook. How do Adams and Wollstonecraft's ideas differ from the ideas in these documents?

_____
_____

© NGSP & HB

SECTION **3** EMERGING EUROPE

3.5 **Nationalism and World War I**

Use with Europe Geography & History, Section 3.5, *in your textbook.*

Reading and Note-Taking **Analyze Cause and Effect**

As you read Section 3.5, use a Cause and Effect Chain to analyze the tensions and events that led to World War I.

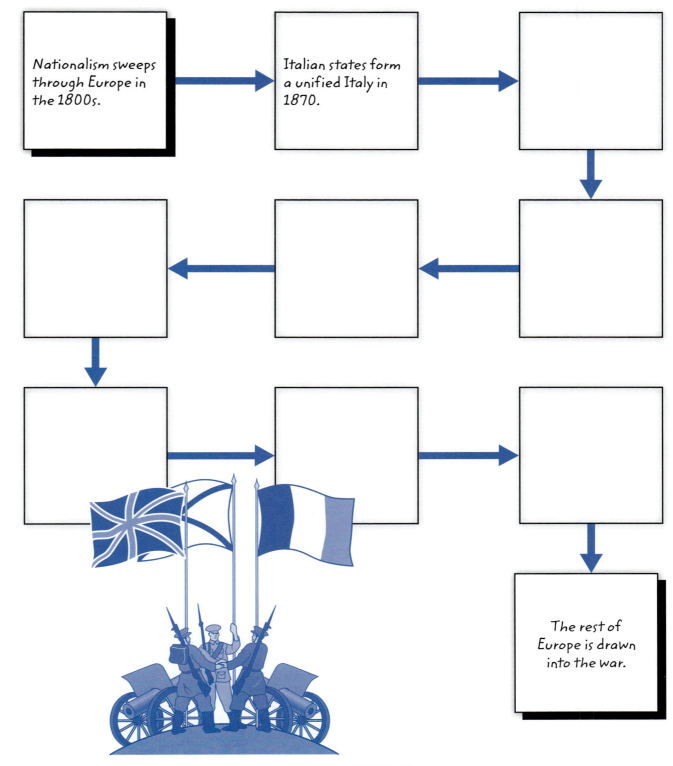

Nationalism sweeps through Europe in the 1800s.

Italian states form a unified Italy in 1870.

The rest of Europe is drawn into the war.

© NGSP & HB

SECTION **3** EMERGING EUROPE

Use with Europe Geography & History, Section 3.5, *in your textbook.*

3.5 Nationalism and World War I

Vocabulary Practice

KEY VOCABULARY

- **nationalism** n., a strong sense of loyalty to one's nation, often with the belief that it is better and more important than other countries
- **trench** n., a long, deep ditch that protects soldiers from enemy gunfire

ACADEMIC VOCABULARY

- **alliance** (uh-LYE-yuhns) n., an agreement among groups or nations to work together toward a common goal

Write a sentence about one *alliance* in your community.

_____

_____

Definition Clues Follow the instructions below for the Key Vocabulary word indicated.

Vocabulary Word: *nationalism*

1. Write the sentence in which the word appears in the section.

   Nationalism is a strong sense of loyalty to

   one's country.

_____

2. Write the definition using your own words.

_____

_____

3. Use the word in a sentence of your own.

_____

_____

_____

4. Why do you think nationalism was a factor leading to World War I?

_____

_____

_____

Vocabulary Word: *trench*

1. Write the sentence in which the word appears in the section.

   Both sides fought from trenches, or long

   ditches that protected soldiers from the

   enemy's gunfire.

2. Write the definition using your own words.

_____

_____

3. Use the word in a sentence of your own.

_____

_____

_____

4. Why do you think soldiers in the Great War used trenches as protection?

_____

_____

_____

Name _____ Class _____ Date _____

Use with Europe Geography & History, Section 3.5, in your textbook.

Go to Interactive Whiteboard GeoActivities at
myNGconnect.com to complete this activity online.

□ **NATIONAL
GEOGRAPHIC**
School Publishing

3.5 NATIONALISM AND WORLD WAR I

Analyze Causes and Effects of World War I

World War I is often called the "Great War" because its destruction was so extensive. Few countries remained neutral, and most of Europe was eventually caught up in the war. Millions of soldiers and civilians were killed. Cities, factories, homes, and farmlands were destroyed.

Many different actions and events combined to bring about the start of the war. Reread Section 3.5 in your textbook. Then use that information to complete the Cause-and-Effect Map below to trace the complicated causes and effects of the war.

Original Cause

First Effect

Second Effect

Intense nationalism → Alliance system →

Assassination of Archduke Ferdinand of Austria →

Communists gain control of Russia →

1. **Make Generalizations** What was the general state of relations among European countries in the early 1900s?

_____
_____
_____

2. **Form and Support Opinions** If Archduke Ferdinand had not been assassinated, would World War I have broken out? Why or why not?

_____
_____
_____

3. **Make Predictions** How did the Treaty of Versailles influence European history in the 1930s and 1940s?

_____
_____
_____

© NGSP & HB

SECTION **3** EMERGING EUROPE

3.6 World War II and the Cold War

Use with Europe Geography & History, Section 3.6, *in your textbook.*

Reading and Note-Taking Sequence Events

As you read about World War II and the Cold War in Section 3.6, use a time line to help you put events in order. Write down important events and the date of each event mentioned.

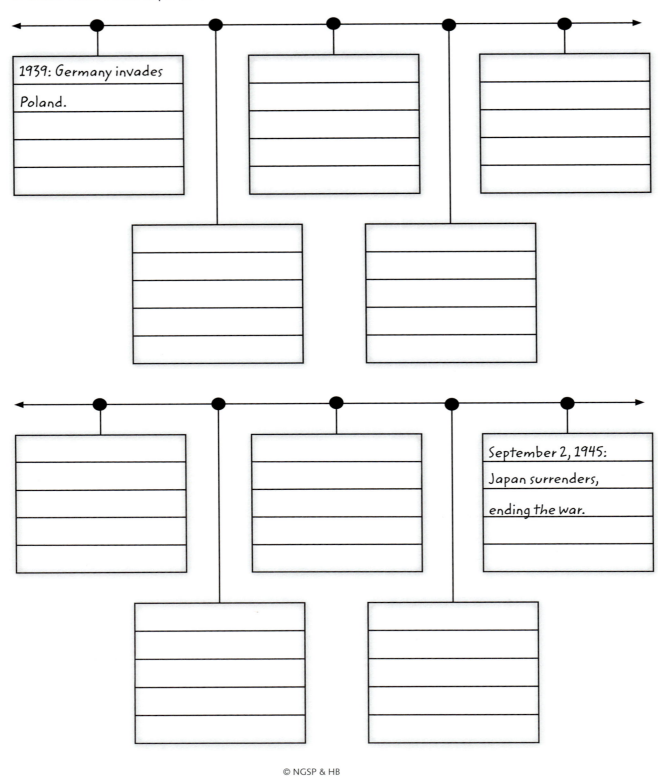

1939: Germany invades Poland.

September 2, 1945: Japan surrenders, ending the war.

© NGSP & HB

SECTION **3** EMERGING EUROPE

Use with Europe Geography & History,
Section 3.6, *in your textbook.*

3.6 **World War II and the Cold War**

Vocabulary Practice

KEY VOCABULARY

- **concentration camp** n., a type of prison where large numbers of civilians are kept during a war and are forced to live in very bad conditions

- **reparations** (rehp-uhr-RAY-shunz) n., money that a country or group that loses a war pays because of the damage, injury, or deaths it has caused

Word Map Complete a Word Map for each Key Vocabulary word. Write the definition and then define the word in your own words. Write the sentence from Section 3.6 that includes the word. Then use the word in a sentence of your own.

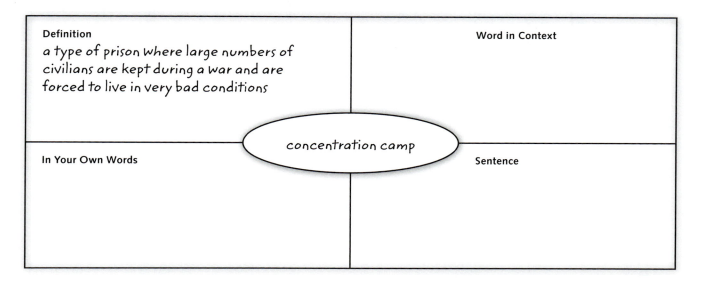

Definition
a type of prison where large numbers of
civilians are kept during a war and are
forced to live in very bad conditions

Word in Context

concentration camp

In Your Own Words

Sentence

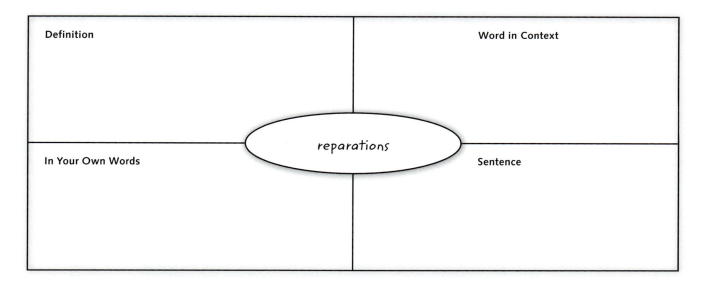

Definition

Word in Context

reparations

In Your Own Words

Sentence

Name _____ Class _____ Date _____

Use with Europe Geography & History, Section 3.6, in your textbook.
Go to Interactive Whiteboard GeoActivities at **myNGconnect.com** *to complete this activity online.*

SECTION **3** EMERGING EUROPE

GeoActivity

3.6 WORLD WAR II AND THE COLD WAR

Analyze the Results of World War II: Berlin

Read the passage and study the map to learn more about Berlin, Germany, after World War II. Then answer the questions.

Dividing Berlin

At the end of World War II in Europe, leaders of the Allies met to decide Germany's future. France, Great Britain, the United States, and the Soviet Union would each govern a piece of the country. Together they would disarm Germany and help the German people rebuild.

The Soviet zone included the city of Berlin. Because the city had been the seat of Hitler's Nazi government, the Allies divided it, too. France, Great Britain, and the United States controlled zones that were known together as West Berlin. The Soviets controlled a large section to the east known as East Berlin. In 1948, when the Western powers united their zones, the Soviet Union launched a blockade of West Berlin. Over many months, planes from the West dropped food and supplies—up to two million tons—and helped West Berliners survive.

The situation worsened over the years with growing tension between East and West Germany. Many people left the East to live in the West. To prevent more people from leaving, the government of East Germany, with help from the Soviets, built a wall in 1961 that split the city and surrounded all of West Berlin. The wall divided the two parts of the city until 1989.

1. **Calculate** Approximately how many miles of wall divided the city of Berlin? How many additional miles of wall surrounded West Berlin?

BERLIN DURING THE COLD WAR

2. **Interpret Maps** What geographic advantage did the Soviet Union and East Germany have over West Berlin?

3. **Make Generalizations** How might the city of Berlin have been seen as a symbol of the Cold War?

© NGSP & HB

SECTION **3** EMERGING EUROPE

Use with Europe Geography & History, Sections 3.1–3.6, *in your textbook.*

3.1–3.6 Review and Assessment

Follow the instructions below to review what you have learned in this section.

Vocabulary Next to each vocabulary word, write the letter of the correct definition.

1. _____ navigation
2. _____ colony
3. _____ factory system
4. _____ radical
5. _____ apartheid
6. _____ nationalism
7. _____ trench
8. _____ reparations

A. money that a country or group that loses a war pays because of the damage, injury, or deaths it has caused

B. a system in which each person works on a small part of a larger project

C. a strong sense of loyalty to one's country

D. the skill of steering modes of transportation and reading maps to get to particular locations

E. an area settled and controlled by a distant country

F. a former social system in South Africa that denied black South Africans their rights

G. a long, deep ditch that protects soldiers from enemy gunfire

H. a person who favors extreme changes in government

Main Ideas Use what you've learned about Europe's history to answer these questions.

9. **Identify** What goods were European explorers interested in finding or obtaining?

_____

_____

10. **Main Ideas** Where did the Industrial Revolution begin and what started it?

_____

_____

11. **Cause and Effect** What were some effects of the Industrial Revolution?

_____

_____

12. **Explain** What three groups composed French society in the late 1700s?

_____

_____

13. **Identify** Who was Napoleon Bonaparte?

_____

_____

14. **Explain** In your own words, what does *natural rights* mean?

_____

_____

15. **Evaluate** What sparked the beginning of World War I?

_____

16. **Describe** What countries made up the Axis Powers in World War II, and who were their opponents?

_____

_____

© NGSP & HB

Focus Skill: Make Inferences

Answer the questions below to make inferences about how Europe developed and extended its influence around the world.

17. Why do you think European leaders wanted to convert people of other lands?

_____

_____

18. Why are quotation marks used for the term *new world*, as in the "new world" that Christopher Columbus uncovered?

_____

_____

19. Why do you think explorers took the time to set up colonies when they landed in a new place?

_____

_____

20. Why would the production of cloth by large machines have made a difference in people's lives in the 1700s?

_____

_____

21. What can you infer about how factory workers felt about industrialization?

_____

_____

22. Why do you think preserving the natural rights of people is important?

_____

_____

23. Why would the Treaty of Versailles after World War I have caused more tension in Europe?

_____

_____

24. Do you think having a class system provides for a peaceful society? Why or why not?

_____

_____

Synthesize: Answer the Essential Question

How did Europe develop and extend its influence around the world? Consider information you have learned about European exploration, colonization, and innovation. Also consider what you have learned about countries and their struggle for power.

_____

_____

_____

_____

_____

© NGSP & HB

SECTION **3** EMERGING EUROPE

3.1–3.6 Standardized Test Practice

Use with Europe Geography & History, *Sections 3.1–3.6, in your textbook.*

Follow the instructions below to practice test-taking on what you've learned from this section.

Multiple Choice Circle the best answer for each question from the options available.

1. For what did Prince Henry of Portugal send explorers to Africa?

   A coal
   B spices
   C gold
   D iron

2. What was the Columbian Exchange?

   A sharing of goods and ideas
   B a monetary rate or amount
   C methods of navigation
   D a way to learn about history

3. Where did the Industrial Revolution start?

   A Portugal
   B Germany
   C France
   D Great Britain

4. How old were the youngest Europeans who worked in factories and mines?

   A 5
   B 12
   C 15
   D 18

5. What did the *Declaration of the Rights of Man and of the Citizen* guarantee?

   A freedom, justice, and voting rights
   B employment, shelter, and food
   C liberty, equality, and property
   D life, happiness, and rights

6. What did Nelson Mandela struggle to end in his country?

   A nationalism
   B communism
   C the system of apartheid
   D the factory system

7. During the 1800s, what effect did nationalism have in Europe?

   A Bismark and Archduke Ferdinand formed an alliance.
   B Nationalism brought about more industry.
   C Nations in Europe united to compete against Asia.
   D Fierce competition among countries led to conflict.

8. Which country surrendered in World War I?

   A Germany
   B Prussia
   C Serbia
   D Austria

9. What started World War II?

   A the Treaty of Versailles
   B Germany invading Poland
   C the assassination of a leader
   D Germany invading France

10. What was the name of the imaginary boundary between Eastern and Western Europe?

   A the Great Depression
   B the Great Wall
   C the Communist Curtain
   D the Iron Curtain

Document-Based Questions

The following excerpt is from **Marie Antoinette: Letter to Her Mother, 1773** from *The World's Story: A History of the World in Story, Song and Art*, published in 1914. Read the excerpt and answer the questions that follow.

> *On Tuesday I had a fête [festival] which I shall never forget all my life. We made our entrance into Paris. As for honors, we received all that we could possibly imagine; but they, though very well in their way, were not what touched me most. What was really affecting was the tenderness and earnestness of the poor people, who, in spite of the taxes with which they are overwhelmed, were transported with joy at seeing us.*

Constructed Response Read each question carefully and write your answer in the space provided.

11. According to the article, how would you describe how Marie Antoinette was feeling?

_____

12. What concern, according to Marie Antoinette, is overwhelming for the poor people?

_____

Extended Response Read each question carefully and write your answer in the space provided.

13. Using what you learned about class structures, what surprises you about Marie Antoinette's acknowledgment of the poor people?

_____

_____

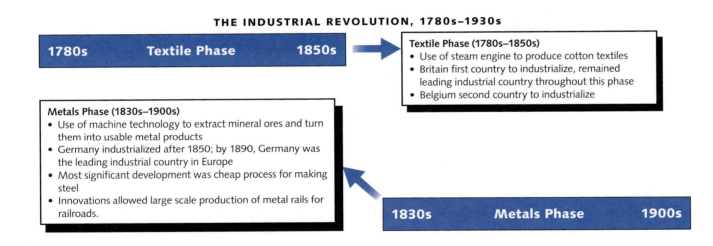

THE INDUSTRIAL REVOLUTION, 1780s–1930s

| 1780s | Textile Phase | 1850s |

Textile Phase (1780s–1850s)
- Use of steam engine to produce cotton textiles
- Britain first country to industrialize, remained leading industrial country throughout this phase
- Belgium second country to industrialize

Metals Phase (1830s–1900s)
- Use of machine technology to extract mineral ores and turn them into usable metal products
- Germany industrialized after 1850; by 1890, Germany was the leading industrial country in Europe
- Most significant development was cheap process for making steel
- Innovations allowed large scale production of metal rails for railroads.

| 1830s | Metals Phase | 1900s |

14. During the Metals Phase, what did machine technology allow people to do?

_____

15. During the Textile Phase, how did Belgium rank in the order of industrialized countries? Given what you know about Britain in Section 3.2, why do you think Belgium ranked as it did?

_____

€UROPE

TODAY • RESOURCE BANK

SECTION **1** CULTURE

Use with Europe Today,
Section 1.1, *in your textbook.*

1.1 Languages and Cultures

Reading and Note-Taking Synthesize Ideas and Details

Use the Concept Cluster to help you take notes about different features of European culture as you read Section 1.1. Write supporting ideas in the ovals and details on the branches. Add ovals and branches if needed. Then synthesize the information in your cluster into a central concept.

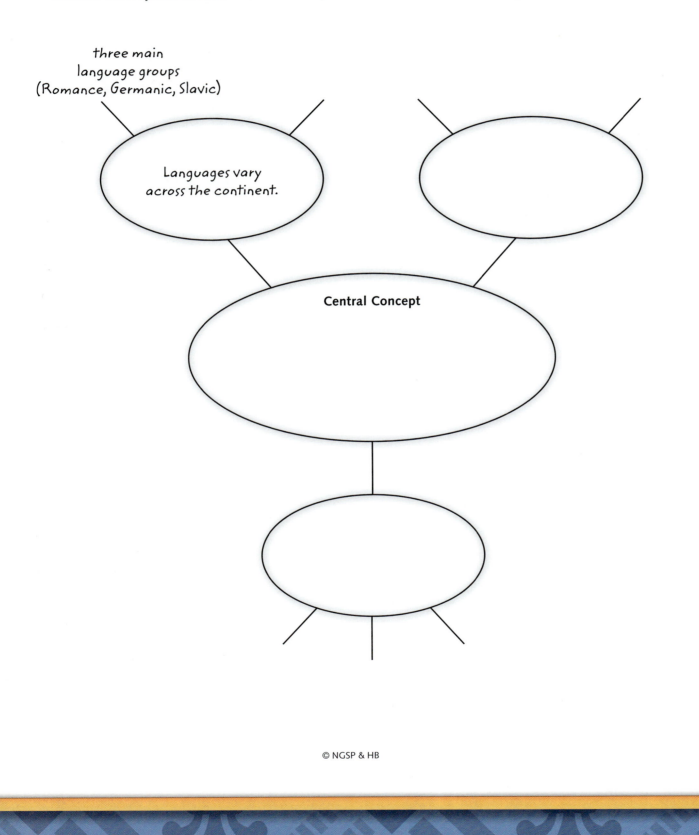

three main
language groups
(Romance, Germanic, Slavic)

Languages vary
across the continent.

Central Concept

Vocabulary Practice

KEY VOCABULARY
- **dialect** (DY-uh-lehkt) n., a regional form or variation of a language
- **heritage** (HEHR-uh-tihj) n., the traditions or beliefs that are part of a country's or people's history

ACADEMIC VOCABULARY
- **cosmopolitan** (kahz-muh-PAH-luh-tuhn) adj., having a broad makeup of cultures and influences from many parts of the world

Write a sentence about city life in Europe using the Academic Vocabulary word *cosmopolitan.*

_____

_____

Word Map Complete a Word Map for the Key Vocabulary words *dialect* and *heritage.* Write the definition for each. To describe what it is like, provide one or two related words or synonyms. Then give one way the word helps us understand European culture.

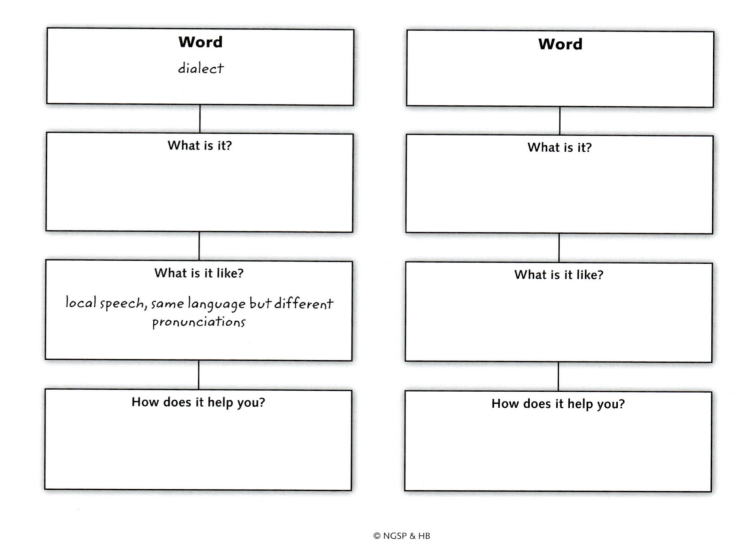

| **Word** |
| dialect |

| **What is it?** |

| **What is it like?** |
| local speech, same language but different pronunciations |

| **How does it help you?** |

| **Word** |

| **What is it?** |

| **What is it like?** |

| **How does it help you?** |

Name _____ Class _____ Date _____

NATIONAL GEOGRAPHIC
School Publishing

SECTION 1 CULTURE
GeoActivity

Use with Europe Today, Section 1.1, in your textbook.
Go to Interactive Whiteboard GeoActivities at
myNGconnect.com *to complete this activity online.*

1.1 LANGUAGES AND CULTURES

Compare Urban Development

European and American cities developed in different ways. Study both maps on this page and answer the questions to compare the layout of a European city with that of an American city.

LONDON, UNITED KINGDOM

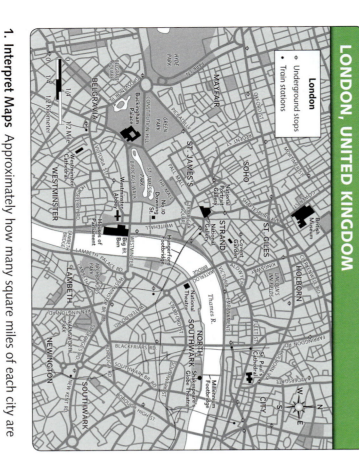

London
○ Underground stops
• Train stations

1. **Interpret Maps** Approximately how many square miles of each city are shown in the map? How can you tell?

_____

CHICAGO, ILLINOIS, UNITED STATES

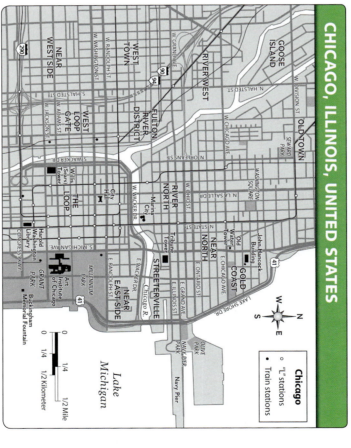

Chicago
○ "L" stations
• Train stations

Lake Michigan

2. **Compare and Contrast** How are the layouts of the cities similar? How are they different?

_____

3. **Draw Conclusions** How might daily life be different for people who live in these cities? Think about the physical and human characteristics of each city as you formulate your response.

_____

© NGSP & HB

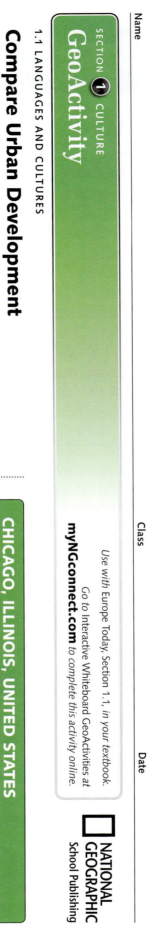

SECTION **1** CULTURE

1.2 Art and Music

Use with Europe Today,
Section 1.2, *in your textbook.*

Reading and Note-Taking Sequence Events

As you read Section 1.2, use the Sequence Chain below to put in order the major developments of European art and music in history. First complete the chain for European art. Then create one for European music. After completing both Sequence Chains, answer the question below.

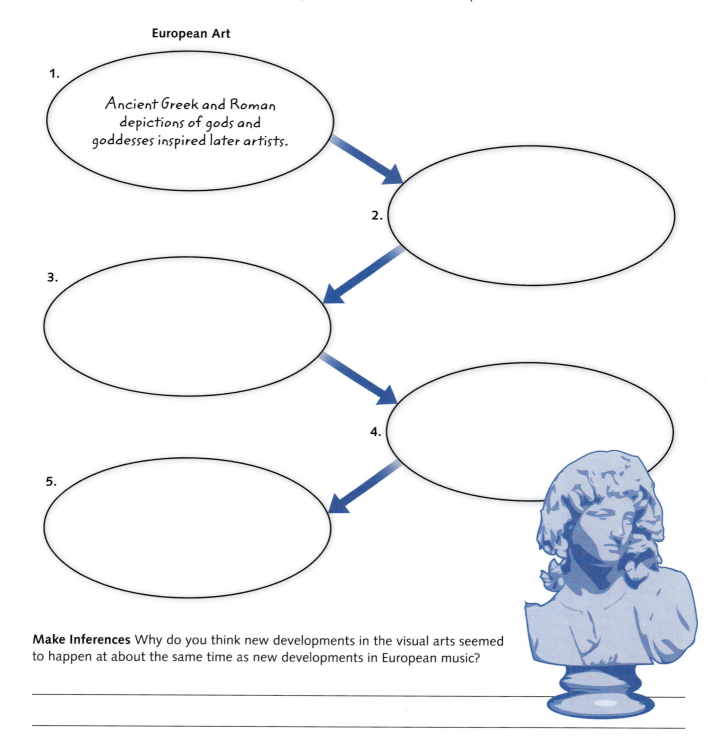

European Art

1. Ancient Greek and Roman depictions of gods and goddesses inspired later artists.

2.

3.

4.

5.

Make Inferences Why do you think new developments in the visual arts seemed to happen at about the same time as new developments in European music?

_____

_____

SECTION **1** CULTURE

1.2 Art and Music

Use with Europe Today, Section 1.2, *in your textbook.*

Vocabulary Practice

KEY VOCABULARY

- **abstract** (ab-STRAKT) adj., expressing ideas with forms or colors without attempting to show something realistically
- **opera** (AH-puh-rah) n., a staged dramatic story in which artistic singing is the main focus
- **perspective** (puhr-SPEHK-tihv) n., a technique used by painters to create the illusion of depth and distance, or three dimensions
- **troubadour** (TROO-buh-dawr) n., a medieval poet-musician

Definition Clues Follow the instructions below for the Key Vocabulary word indicated.

Vocabulary Word: <u>abstract</u>

1. Write the sentence in which the word appears in the lesson.

<u>These modern artists often worked in an abstract style, which means they emphasized form and</u>

<u>color over realism.</u>

2. Write a definition of the word using your own words.

_____

3. Write a sentence using the word correctly.

_____

4. Now that you understand the word *abstract*, what would you expect an abstract painting to look like? Would it include *perspective*? Why or why not?

_____

_____

Vocabulary Words: <u>troubadour</u> <u>opera</u>

5. Describe how singers have been a part of European music since ancient Greece and Rome. Use the Key Vocabulary words *troubadour* and *opera* in your description.

_____

_____

_____

SECTION **1** CULTURE

GeoActivity

1.2 ART AND MUSIC

Recognize Architectural Movements

Use with Europe Today, Section 1.2, in your textbook.

Go to Interactive Whiteboard GeoActivities at
myNGconnect.com *to complete this activity online.*

NATIONAL GEOGRAPHIC
School Publishing

Like Europe's art and music, European architecture has experienced several different movements. Read about the characteristics of some of these movements. Then identify the movement for each famous building pictured.

Styles Develop Through Time

Gothic This movement developed in the 12th and 13th centuries. Engineers found new methods and tools to build taller, more impressive structures. Pointed arches and stained-glass windows allowed buildings to fill with light in an era long before electricity.

Neoclassical This style of architecture became very popular in the 18th and 19th centuries. Architects wanted to bring back the simpler, classical styles from ancient Greece and Rome. Many of the major buildings from this era have columns, triangular pediments, and round domes.

Modernist This movement developed in the early 20th century. Many architects stripped away unnecessary ornamentation, following the idea that "form follows function," or that a building's appearance should be designed around its purpose. Architects took advantage of new materials, creating buildings out of steel, glass, and reinforced concrete.

Analyze Visuals What movement does each building at right belong to? Explain why you think so.

1. Bauhaus (Dessau, Germany)

Movement: _____

Explanation: _____
_____
_____

2. Canterbury Cathedral (Canterbury, United Kingdom)

Movement: _____

Explanation: _____
_____
_____

3. Panthéon (Paris, France)

Movement: _____

Explanation: _____
_____
_____

SECTION **1** CULTURE

1.3 Europe's Literary Heritage

Use with Europe Today, Section 1.3, *in your textbook.*

Reading and Note-Taking Find Main Idea and Details

As you read Section 1.3, use a Main Idea Diagram to help you take notes about the history of European literature.

Main Idea:

European literature, including poetry, plays, and novels, has reflected new ways of thinking over the centuries.

Detail:

Detail:

Detail:

Detail:

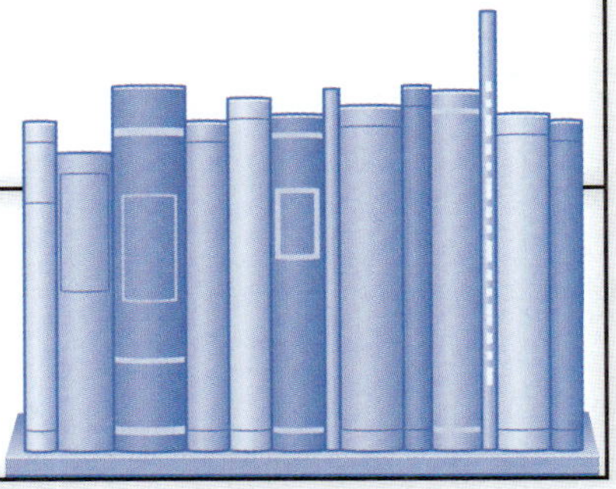

Detail:

Detail:

© NGSP & HB

Name _____ Class _____ Date _____

Vocabulary Practice

KEY VOCABULARY

- **epic** (EH-pik) **poem** n., a long poem that tells the adventures of a legendary or historical hero
- **genre** (ZHAHN-ruh) n., an artistic, musical, or literary form or category
- **novel** (NAH-vuhl) n., a long work of fiction that includes complex characters and a plot

Compare/Contrast Paragraph Write a short paragraph comparing the definition and description of the epic poem with the novel, using at least one example of each from Section 1.3. Use all of the Key Vocabulary words in your paragraph. Begin your paragraph with a clear topic sentence and end with a concluding sentence that summarizes your main points.

Topic Sentence: _____

_____

_____

_____

_____

_____

_____

_____

_____

_____

_____

_____

Summarizing Sentence: _____

_____

_____

_____

Name _____ Class _____ Date _____

Use with Europe Today, Section 1.3, in your textbook.

Go to Interactive Whiteboard GeoActivities at
myNGconnect.com *to complete this activity online.*

SECTION ① CULTURE
GeoActivity

1.3 EUROPE'S LITERARY HERITAGE

Analyze Primary Sources: Romantic Writing

Romanticism was an artistic movement of dramatic expression and intense emotion. Many Romantic works, whether painting, music, or literature, also celebrated the glory of nature. Read the excerpts from John Keats and Mary Shelley, two significant writers from the Romantic period.

Document 1: John Keats on the Proper Role of Poetry

Poetry should be great and unobtrusive [not showy], a thing which enters into one's soul, and does not startle it or amaze it with itself, but with its subject. How beautiful are the retired [quiet and secluded] flowers! How would they lose their beauty were they to throng into the highway crying out, 'Admire me I am a violet! Dote upon me I am a primrose!'

—from a letter to John Hamilton Reynolds, February 3, 1818

Keats was already a celebrated poet when he died at the young age of 25.

Document 2: Mary Shelley on the Power of Poets

What is a poet? Is he not that which wakens melody in the silent chords of the human heart? A light which arrays [displays] in splendor things and thoughts which else were dim in the shadow of their own significance. . . . He is the mirror of nature, reflecting her back ten thousand times more lovely; what then must not his power be, when he adds beauty to the most perfect thing in nature—even Love.

—from "The Loves of the Poets,"
Westminster Review, October 1829

Shelley was the author of the gothic novel *Frankenstein*.

1. **Analyze Primary Sources** What does Keats suggest should be the major focus of a poem? In his view, how should a poem startle and amaze the reader?

_____
_____
_____

2. **Compare and Contrast** Does Shelley agree or disagree with Keats about the role of the poet? Explain.

_____
_____
_____

3. **Synthesize** Romantic painters had ideas about painting that were similar to Keats and Shelley's views of poetry. Would you expect Romantic paintings to be abstract or realistic? Why?

_____
_____
_____

© NGSP & HB

Name _____ Class _____ Date _____

1.4 Cuisines of Europe

Use with Europe Today, Section 1.4, *in your textbook.*

Reading and Note-Taking Categorize Cuisines

Use the Classification Chart below to help you categorize different types of European cuisines. List a European country or area in each box. Then add two or three details, staple ingredients, or traditional dishes from Section 1.4 for each type of cuisine specific to the areas listed.

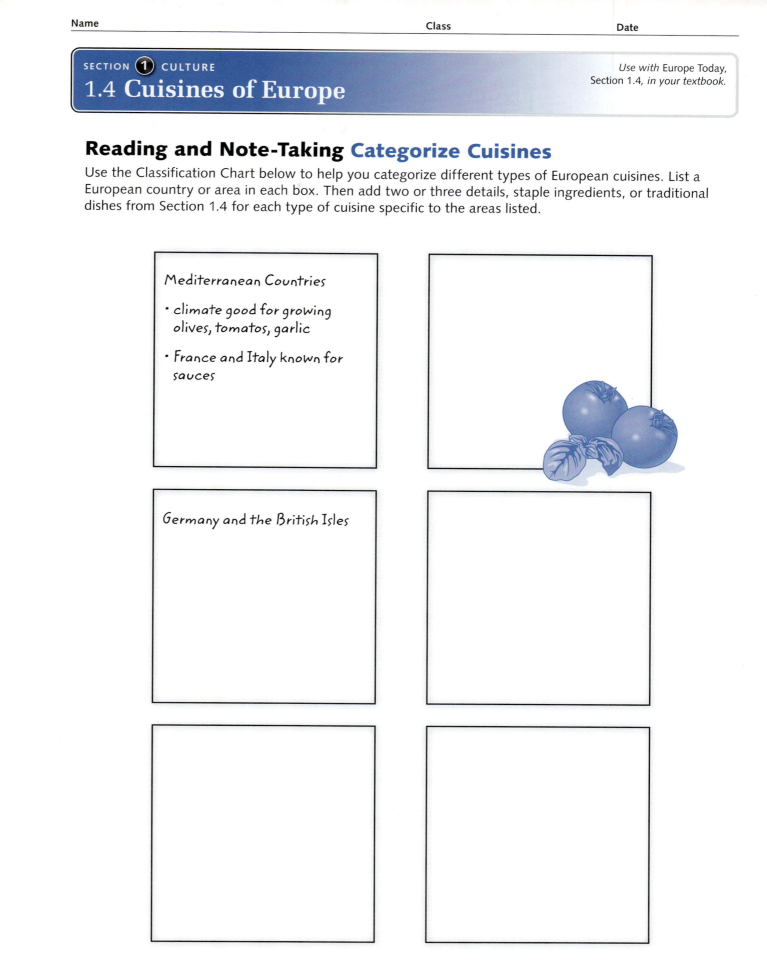

Mediterranean Countries

• climate good for growing olives, tomatos, garlic

• France and Italy known for sauces

Germany and the British Isles

© NGSP & HB

SECTION **1** CULTURE

1.4 Cuisines of Europe

Use with Europe Today, *Section 1.4, in your textbook.*

Vocabulary Practice

KEY VOCABULARY

• **cuisine** (kwih-ZEEN) n., the traditional cooking style of a culture

• **staple** (STAY-puhl) n., a basic food that is part of people's diets

Comparison Chart Complete the Y-Chart to compare the meaning of the Key Vocabulary words *cuisine* and *staples*. Then write how the two words are related.

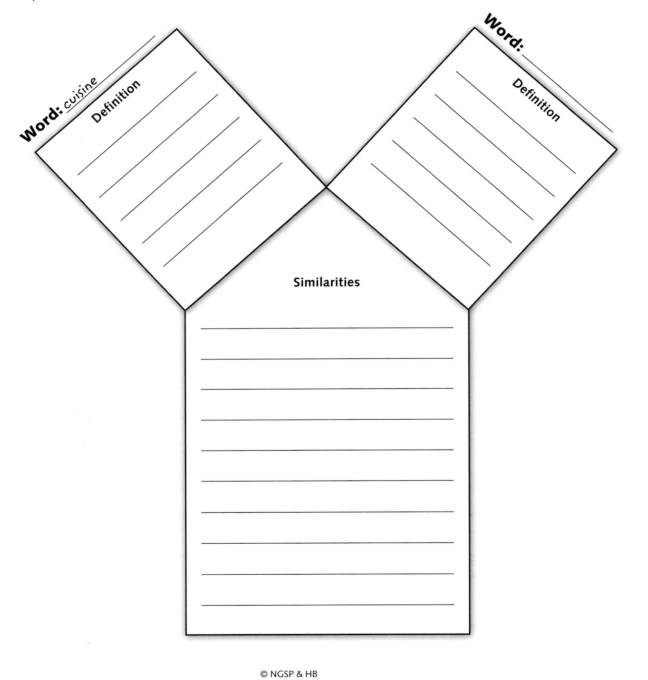

Word: *cuisine*

Definition

Word: _____

Definition

Similarities

SECTION 1 CULTURE
GeoActivity

Use with Europe Today, Section 1.4, in your textbook.

Go to Interactive Whiteboard GeoActivities at
myNGconnect.com *to complete this activity online.*

☐ **NATIONAL GEOGRAPHIC**
School Publishing

1.4 CUISINES OF EUROPE

Graph Olive Oil Production Rates

Olive oil has been a staple of diets in Spain, Italy, Greece, and other Mediterranean countries for centuries. In the last two decades, people have learned about the health benefits of olive oil, and demand for it has grown worldwide. As a result, olive oil production has begun in other countries, including Chile, Australia, and the United States. The chart below contains data on olive oil production in several old- and new-producing countries. Study the chart and then answer the questions.

Olive Oil Production

COUNTRY (new/old producer)	1980s AVERAGE ANNUAL PRODUCTION (metric tons)	1990s AVERAGE ANNUAL PRODUCTION (metric tons)	2000s AVERAGE ANNUAL PRODUCTION (metric tons)	PERCENTAGE OF INCREASE (1980s-2000s)
Australia (new)	113	88	1,869	
Chile (new)	938	1,020	1,771	
Greece (old)	293,095	348,991	353,948	
Italy (old)	554,232	529,328	605,132	
Spain (old)	495,098	683,879	1,100,668	
United States (new)	550	627	1,361	

Source: Food and Agriculture Organization of the United Nations, Statistics Division

1. **Analyze Data** In which decade did overall olive oil production seem to increase the most? Why did this happen?

_____

2. **Calculate** Use the data in the chart to calculate the percentage by which olive oil production increased in each country between the 1980s and the 2000s. (Subtract the 1980s amount from the 2000s amount. Divide by the 1980s amount and multiply by 100.) Write the percentages in the fifth column of the chart.

3. **Create Graphs** On the grid below, create a bar graph that shows the percentage of increase for each country.

Olive Oil Production

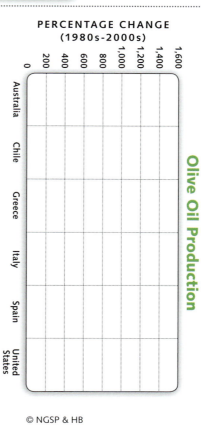

PERCENTAGE CHANGE
(1980s-2000s)

1,600	
1,400	
1,200	
1,000	
800	
600	
400	
200	
0	

Australia Chile Greece Italy Spain United States

4. **Interpret Graphs** Which country had the greatest percentage of increase? Which country had the lowest percentage of increase? What trend do you notice about old-producing countries and new-producing countries?

5. **Synthesize** How might old producers benefit from the increased popularity of olive oil? What drawbacks might there be?

_____
_____
_____

© NGSP & HB

SECTION **1** CULTURE

1.1–1.4 Review and Assessment

Use with Europe Today, Sections 1.1–1.4, *in your textbook.*

Follow the instructions below to review what you have learned in this section.

Vocabulary Next to each vocabulary word, write the letter of the correct definition.

1. ____ dialect

2. ____ heritage

3. ____ perspective

4. ____ troubadour

5. ____ genre

6. ____ novel

7. ____ staple

8. ____ cuisine

A. a medieval poet-musician

B. a long work of fiction that includes complex characters and a plot

C. a technique used by painters to create the illusion of depth and distance, or three dimensions

D. the traditional cooking style of a culture

E. a regional form or variation of a language

F. an artistic, musical, or literary form or category

G. a basic food that is part of people's diets

H. the traditions or beliefs that are part of a country's or people's history

Main Ideas Use what you've learned about Europe's culture to answer these questions.

9. **Compare** Is population density in Europe higher or lower than in the United States?

_____

_____

10. **Identify** In which part of Europe do most people speak Slavic languages?

_____

_____

11. **Make Generalizations** What subject matter was most commonly shown in European art from ancient times through the Renaissance?

_____

_____

12. **Sequence** Which musical period came first: Classical, Romantic, or Baroque?

_____

_____

13. **Analyze Cause and Effect** What effect did the invention of the printing press have on literature?

_____

_____

14. **Synthesize** What did writers and visual artists in the early 20th century have in common?

_____

_____

15. **Compare** How do cuisines in Europe's cooler regions compare with those in warmer countries?

_____

_____

16. **Form and Support Opinions** Do you think seafood is a more important part of the cuisine in Portugal or in the Czech Republic? Why?

_____

_____

Focus Skill: Draw Conclusions

Answer the questions below to draw conclusions about European culture.

17. What effect have Europe's dozens of different languages had on cultures across the continent?

_____

_____

18. What effect might having multiple official languages have on the culture of a small country such as Belgium?

_____

_____

19. Why might most Renaissance paintings appear more realistic than earlier paintings from the Middle Ages?

_____

_____

20. Baroque period music is known for its use of complex forms. What traits would you expect to find in Baroque painting?

_____

_____

21. In what ways did the modern literature of the 20th century differ from 19th-century literature?

_____

_____

22. How were the works of Jane Austen, Charles Dickens, and Henrik Ibsen different from the works of earlier writers?

_____

_____

23. Many of the staple ingredients in European dishes were originally introduced from other regions. What does this indicate about Europe's local cultures and traditions?

_____

_____

24. What geographic factors explain why bread is so important to Ukrainian cuisine?

_____

_____

Synthesize: Answer the Essential Question

How is the diversity of Europe reflected in its cultural achievements? Recall what you have read about Europe's large number of separate countries, with distinct languages and cultures. Consider as well the artistic and literary movements that have spread across borders, and the other ways people in these countries have influenced one another.

_____

_____

_____

_____

_____

_____

_____

© NGSP & HB

SECTION **1** CULTURE

1.1–1.4 **Standardized Test Practice**

Use with Europe Today,
Sections 1.1–1.4, *in your textbook.*

Follow the instructions below to practice test-taking on what you've learned from this section.

Multiple Choice Circle the best answer for each question from the options available.

1. Which of the following is a Germanic language?

   A French
   B Dutch
   C Spanish
   D Polish

2. What is Europe's fastest-growing religion?

   A Judaism
   B Christianity
   C Hinduism
   D Islam

3. What did the Impressionist painters attempt to do?

   A emphasize form and color over realism
   B present religious subjects as two-dimensional figures
   C use light and color to capture a moment
   D portray gods and goddesses with realistic human forms

4. During which period did many artists begin working in an abstract style?

   A mid-17th century
   B late 18th century
   C mid-19th century
   D early 20th century

5. Opera began during which period?

   A Romantic
   B Renaissance
   C Baroque
   D Middle Ages

6. What book is considered to be the first modern novel?

   A *The Iliad*
   B *Hamlet*
   C *Don Quixote*
   D *Sense and Sensibility*

7. What was the name of the Greek poet who wrote *The Iliad* and *The Odyssey*?

   A Homer
   B Voltaire
   C Virgil
   D Dante

8. What was the effect of the two world wars on 20th-century literature?

   A Writers explored the rights of the individual.
   B Writers reflected the sense that life was unpredictable.
   C Writers criticized the traditional roles of husbands and wives.
   D Writers such as Goethe emphasized emotion and nature.

9. The climate in Mediterranean countries allows them to develop which foods?

   A wheat and other grains
   B root vegetables such as turnips and beets
   C herring, deer, and elk meat
   D olives, tomatoes, and garlic

10. What is *goulash*?

    A thin French pancake
    B Hungarian beef stew
    C Ukrainian type of bread
    D Russian beet soup

Document-Based Questions

The following excerpt is from the website of the United Nations Educational, Scientific, and Cultural Organization, which recognizes the value of the Mediterranean diet as part of humanity's cultural heritage. Read the paragraph and answer the questions that follow.

The Mediterranean diet is characterized by a nutritional model that has remained constant over time and space, consisting mainly of olive oil, cereals, fresh or dried fruit and vegetables, a moderate amount of fish, dairy, and meat, and many condiments and spices. . . . It promotes social interaction, since communal [shared] meals are the cornerstone of social customs and festive events.

Constructed Response Read each question carefully and write your answer in the space provided.

11. According to the passage, how does the Mediterranean diet promote social interaction?

_____

12. Are fruits and vegetables a larger or a smaller part of the Mediterranean diet than meat and dairy? What term or phrase does the passage use to indicate this?

_____

Extended Response Read each question carefully and write your answer in the space provided.

13. What features distinguish the Mediterranean diet from other European cuisines? What characteristics might they share?

_____

_____

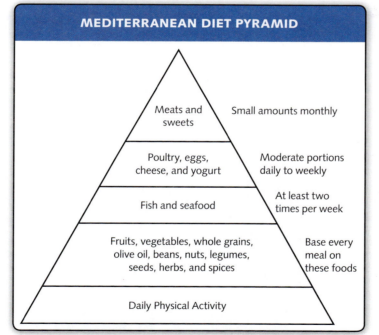

MEDITERRANEAN DIET PYRAMID

Meats and sweets — Small amounts monthly

Poultry, eggs, cheese, and yogurt — Moderate portions daily to weekly

Fish and seafood — At least two times per week

Fruits, vegetables, whole grains, olive oil, beans, nuts, legumes, seeds, herbs, and spices — Base every meal on these foods

Daily Physical Activity

Sources: 2009 Oldways Preservation & Exchange Trust; MayoClinic.com

14. According to the diagram, which animal products are healthiest? Which should be eaten on only an occasional basis?

_____

_____

15. Based on what you have read, what advantages might there be to the increasing popularity of the Mediterranean diet around the world? What disadvantages might there be as a result?

_____

_____

SECTION 2 GOVERNMENT & ECONOMICS
2.1 The European Union

Use with Europe Today,
Section 2.1, *in your textbook.*

Reading and Note-Taking Outline and Take Notes

Use an outline to help you take notes about the European Union as you read Section 2.1.

The Origins and Structure of the European Union

I. The European Union's Origins _____

   A. Organization for European Economic Cooperation forms in 1948 ____

   B. The Common Market: Western Europe sought closer economic ties __

   C. Treaty of Maastricht: _____

II. _____

   A. _____

   B. _____

   C. _____

III. _____

   A. _____

   B. _____

   C. _____

Flag of the European Union

SECTION 2 GOVERNMENT & ECONOMICS

2.1 The European Union

Use with Europe Today, Section 2.1, in your textbook.

Vocabulary Practice

KEY VOCABULARY

- **euro** (YUR-oh) n., the basic monetary unit of most European Union countries
- **sovereignty** (SAHV-uh-ruhn-tee) n., a state's control over its own affairs, free from outside control
- **tariff** (TEHR-uhf) n., a tax paid on imports and exports

ACADEMIC VOCABULARY

- **currency** (KUHR-uhn-see) n., money used as a medium of exchange

Write a sentence using the Academic Vocabulary word *currency*.

_____

_____

Words in Context Follow the instructions below for using the Key Vocabulary in writing.

1. Describe how high *tariffs* could create barriers against international trade.

_____

_____

2. Explain how concerns about *sovereignty* have led some countries to avoid joining the European Union.

_____

_____

3. Write a sentence using the word *euro*.

_____

_____

4. Briefly explain how adopting the *euro* could affect a country's *sovereignty*.

_____

_____

5. Describe the effect of the Common Market on *tariffs*.

_____

_____

6. Use two of the three vocabulary words in a sentence.

_____

_____

Name _____ Class _____ Date _____

NATIONAL
GEOGRAPHIC
School Publishing

SECTION **2** GOVERNMENT & ECONOMICS

GeoActivity

2.1 THE EUROPEAN UNION

Categorize Fundamental Rights

In 2000, the European Union wrote its Charter of Fundamental Rights. This charter guarantees basic human rights for all EU citizens. The charter divides these rights among several categories:

a. Dignity (being worthy of respect)

b. Freedoms (actions that cannot be restricted by any authority)

c. Equality (having the same rights as everyone else)

d. Solidarity (rights of workers)

e. Citizens' Rights (rights a person has under a government)

f. Justice (using laws fairly to judge or punish people)

Read the excerpts from the charter and match them to the appropriate category. Write the letter of the category on the line. Then answer the questions.

1. ____ Equality between men and women must be ensured in all areas, including employment, work, and pay.

2. ____ Everyone is entitled to a fair and public hearing within a reasonable time by an independent and impartial tribunal previously established by law. Everyone shall have the possibility of being advised, defended, and represented.

3. ____ Human dignity is inviolable. It must be respected and protected.

4. ____ Every citizen of the Union has the right to vote and to stand as a candidate at municipal elections in the Member State in which he or she resides under the same conditions as nationals of that State.

5. ____ Every worker has the right to working conditions which respect his or her health, safety, and dignity.

6. ____ Everyone has the right to freedom of expression. This right shall include freedom to hold opinions and to receive and impart information and ideas without interference by public authority and regardless of frontiers.

7. **Make Inferences** The EU Court of Justice was created to monitor EU laws, including the Charter of Fundamental Rights. However, these laws apply to many countries, each with its own government and justice system. What problems might arise in enforcing these laws?

_____

_____

_____

_____

8. **Draw Conclusions** Most EU countries already had laws defending the rights guaranteed by the charter. What benefits would there be for the EU as an organization to officially protect these rights?

_____

_____

_____

_____

Use with Europe Today, Section 2.1, in your textbook.

Go to Interactive Whiteboard GeoActivities at **myNGconnect.com** *to complete this activity online.*

SECTION **2** GOVERNMENT & ECONOMICS

Use with Europe Today,
Section 2.2, *in your textbook.*

2.2 The Impact of the Euro

Reading and Note-Taking **Draw Conclusions**

As you read Section 2.2, use the chart below to help you note the different effects that the euro has had on European economics and politics. Then draw a conclusion about the overall consequences of the region's transition to a shared currency.

Topic	Effect	Outcome
Tourism		Travel is easier and less expensive.
Conducting business		

Draw Conclusions

SECTION ② GOVERNMENT & ECONOMICS

2.2 The Impact of the Euro

Use with Europe Today,
Section 2.2, *in your textbook.*

Vocabulary Practice

KEY VOCABULARY

• **eurozone** (YUR-oh-zohn) n., the group of countries using the euro

• **exchange** (ehks-CHAYNJ) v., to convert money from one currency to another

ACADEMIC VOCABULARY

• **consumer** (kuhn-SOO-muhr) n., one who buys or uses economic goods

Use the Academic Vocabulary word *consumer* in a sentence.

_____

_____

Travel Blog Write a blog entry about planning a trip to France, Germany, and Great Britain and the preparations you need to make for your trip. Consider any information you would need to know before leaving, such as money to take and items to pack. Use both Key Vocabulary words in your blog.

Blog title and date: _____

Blog entry: _____

_____

_____

_____

_____

_____

_____

_____

_____

_____

_____

_____

_____

© NGSP & HB

BLACK-AND-WHITE COPY READY

GeoActivity

2.2 THE IMPACT OF THE EURO

Use with Europe Today, Section 2.2, in your textbook.

Go to Interactive Whiteboard GeoActivities at **myNGconnect.com** *to complete this activity online.*

□ NATIONAL GEOGRAPHIC School Publishing

Graph the Economic Impact of the Euro

How have the European Union and the euro affected economies across Europe? Have they benefited poorer countries, such as Ireland and Slovenia, as much as they have benefited wealthier countries, such as Germany and France? The first chart shows standards of living in Germany and Ireland as indicated by GDP per capita. The second chart shows the percentage of each country's GDP made up of exported goods and services. Study the charts and answer the questions that follow.

GDP Per Capita
(current U.S. dollars)

	GERMANY	IRELAND
1990	$21,600	$13,600
1995	$30,900	$18,600
2000	$23,100	$23,400
2005	$33,800	$48,500
2010	$40,700	$51,000

Source: World Bank

Exports of Goods and Services
(percentage of GDP)

	GERMANY	IRELAND
1990	25%	57%
1995	24%	76%
2000	33%	98%
2005	41%	82%
2010	41%	89%

Source: World Bank

1. Create Graphs Use the information on each country from the charts to create a line-bar graph at right. Use the values on the left y-axis to plot lines that represent GDP per capita. Use the values on the right y-axis to plot bars that represent exports as a percentage of total GDP. Choose different colors for each country and category. Identify them in the legend.

2. Analyze Data Was trade an important part of the Irish economy before the country adopted the euro in 1999? How can you tell? How did this change after switching to the euro?

_____
_____
_____
_____

Impact of the Euro

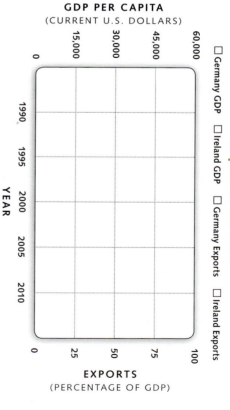

☐ Germany GDP ☐ Ireland GDP ☐ Germany Exports ☐ Ireland Exports

GDP PER CAPITA (CURRENT U.S. DOLLARS)

EXPORTS (PERCENTAGE OF GDP)

3. Make Inferences During the 1990s Germany was adapting to the reunification of the country and its economy was unsteady. How could the adoption of the euro have helped during this period?

_____
_____
_____

4. Make Predictions Several countries in Eastern Europe that have lower standards of living have adopted the euro in recent years. Others are scheduled to do so over the next decade. What would you expect the effect to be for their economies?

_____
_____
_____
_____

© NGSP & HB

SECTION ② GOVERNMENT & ECONOMICS

2.3 Democracy in Eastern Europe

Use with Europe Today, Section 2.3, *in your textbook.*

Reading and Note-Taking Summarize Governments and Economies

As you read Section 2.3, use the chart below to help you summarize details about the governments and economies of countries in Eastern Europe since the fall of the Soviet Union.

Government Transitions	Economic Transitions
• After rebellions against the Communist government, countries declared independence. • Poland, Hungary, and the Czech Republic developed stable democracies.	

Summary

SECTION **2** GOVERNMENT & ECONOMICS

2.3 Democracy in Eastern Europe

Use with Europe Today,
Section 2.3, *in your textbook.*

Vocabulary Practice

KEY VOCABULARY

- **democratization** (dih-MAH-kruh-ty-zay-shuhn) n., the process of becoming a government in which the people hold political authority

- **privatization** (PRY-vuh-ty-zay-shuhn) n., the process of transferring control of government-owned businesses to private owners

Definition Chart Complete a Definition Chart for each of the Key Vocabulary words.

Word	democratization
Definition	the process of becoming a government in which the people hold political authority
In Your Own Words	
Use in Sentence	

Word	privatization
Definition	
In Your Own Words	
Use in Sentence	

© NGSP & HB

Name _____ Class _____ Date _____

Use with Europe Today Section 2.3, *in your textbook.*

Go *to Interactive Whiteboard GeoActivities at*
myNGconnect.com *to complete this activity online.*

NATIONAL GEOGRAPHIC School Publishing

SECTION ② GOVERNMENT & ECONOMICS

GeoActivity

2.3 DEMOCRACY IN EASTERN EUROPE

Research Reform Movements

The collapse of communism spread rapidly across Eastern Europe in the late 1980s, but the pace of democratization in the region varied. Some countries experienced smooth transitions while others ended up in civil wars. Research one of these countries to learn more about its fate after communism. Use the steps in the Project Organizer to prepare a report on your country.

Step 1 Select a Country Choose one of the following countries to research:

- Czechoslovakia
- Poland
- Ukraine
- Hungary
- Romania
- Yugoslavia

Step 2 Identify Your Sources Use the library or the research links at **Connect to NG** to find at least one primary source and one secondary source on your country.

Step 3 Research and Take Notes Read your sources to learn about events in your country during and after the fall of communism. Use the research questions in the Project Organizer to guide your research. Take notes on a separate sheet of paper.

Step 4 Identify Your Main Idea Write a sentence that summarizes what you learned about the fate of your country after the fall of communism. For example, a sentence about Germany might be, "After years as a symbol of tension between East and West, the Berlin Wall came down peacefully."

Step 5 Present Your Findings Now create a presentation about your country so you can share what you learned with your classmates. Use the presentation formats and tips at right to choose how you will present your findings.

Project Organizer

Step 1 Select a Country

Write the name of your country here: _____

Step 2 Identify Your Sources

Primary Source: _____

Secondary Source: _____

Step 3 Research and Take Notes

Research Questions:

- What event or political movement helped defeat communism?
- Who were key leaders of the event or movement?
- Was the process of democratization easy or difficult for the people in the country?
- What events made it easy? What events made it difficult?
- Did the country survive the process, or did it dissolve into new countries?
- What is the state of the original country or new countries today?

Step 4 Identify Your Main Idea

_____

Step 5 Present Your Findings

Essay
- Create an outline.
- Use a formal tone.
- Read your essay to the class.

Blog
- Think about your audience.
- Use a less formal tone.
- Post your blog entries.

Presentation
- Write clear, simple bullet points.
- Use appropriate visuals.
- Project your presentation in class.

Video/Photo Essay
- Stay under 5 minutes.
- Include charts or graphs.
- Play your visual essay for the class.

Presentation Format: _____

© NGSP & HB

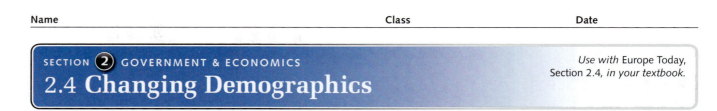

SECTION **2** GOVERNMENT & ECONOMICS
2.4 Changing Demographics

Use with Europe Today, Section 2.4, in your textbook.

Reading and Note-Taking Analyze Cause and Effect

Use the Cause-and-Effect Web below to help you take notes about the different factors contributing to Europe's changing demographics. Then synthesize the information to answer the question below.

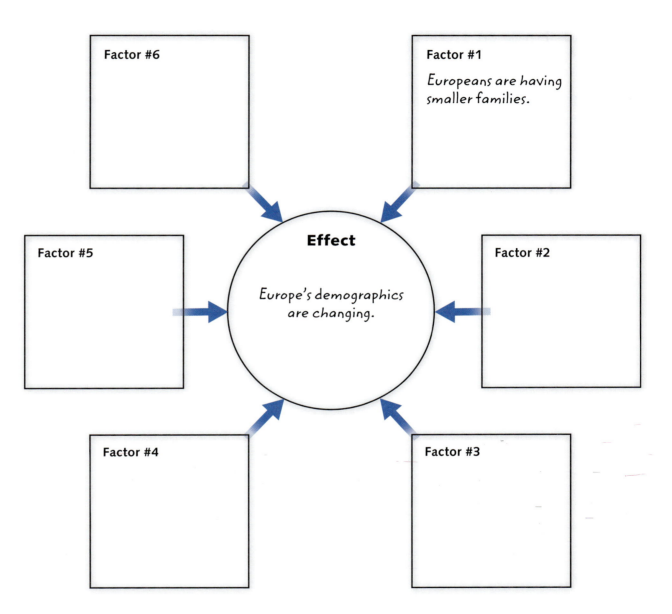

Synthesize How has Europe's changing demographics challenged the region? How has it benefited the region?

_____

_____

SECTION **2** GOVERNMENT & ECONOMICS

2.4 Changing Demographics

Use with Europe Today,
Section 2.4, *in your textbook.*

Vocabulary Practice

KEY VOCABULARY

- **aging population** n., a trend that occurs as the average age of a population rises
- **assimilate** (uh-SIH-muh-layt), v., to be absorbed into a society's culture
- **demographic** (deh-muh-GRA-fiks), n., the characteristic of a human population, such as age, income, and education

Related Idea Web Inside each circle, write one of the Key Vocabulary words and its definition. On the lines connecting the circles, write a sentence about how the two words are related.

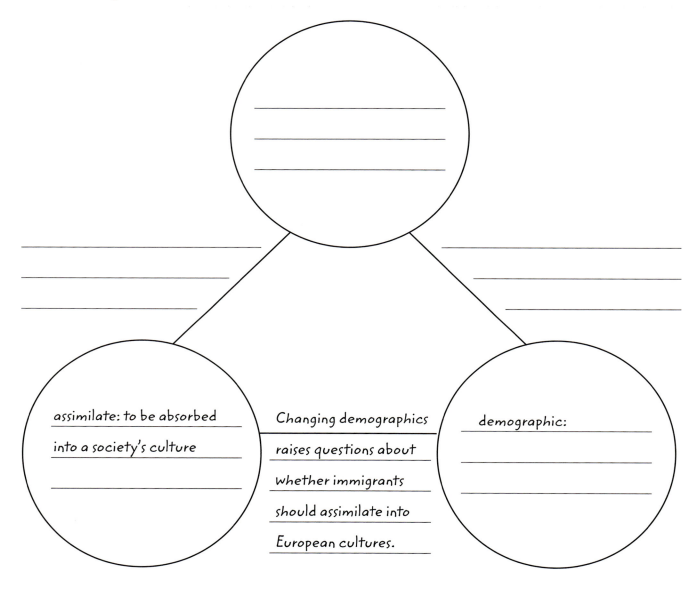

assimilate: to be absorbed
into a society's culture

Changing demographics
raises questions about
whether immigrants
should assimilate into
European cultures.

demographic:

SECTION **2** GOVERNMENT & ECONOMICS

GeoActivity

2.4 GLOBAL ISSUES: CHANGING DEMOGRAPHICS

Compare London's Immigrant Populations

Use with Europe Today, Section 2.4, in your textbook.

Go to Interactive Whiteboard GeoActivities at
myNGconnect.com *to complete this activity online.*

☐ NATIONAL
GEOGRAPHIC
School Publishing

London, the capital of the United Kingdom, is one of the most diverse cities in the world. More than 300 different languages are spoken there. What has brought so many people to London? Read about four of London's largest immigrant populations. Then fill in the chart with push and pull factors that influence immigration for each population.

A Global Destination

People from India have lived in London as far back as the 1600s. After India gained independence in 1947, many people relocated to London. Labor shortages enabled many Indians to find work in the textile trade. In the 1960s, the British National Health Service began recruiting doctors from India, and many Indians continue to work in London's health care industry today.

The division of British India also resulted in great upheavals in the area that would become Bangladesh. In the following decades, some Bangladeshis left their homes and started new lives as textile workers in London. Other Bangladeshis opened restaurants, cafes, and shops in London's Brick Lane area, providing jobs for newer immigrants.

For centuries, Ireland's people have left their homes to escape poverty and find work in British cities. The labor shortages that led many Indians to London in the 1950s and 1960s also attracted many Irish immigrants, who found jobs as nurses and construction workers. Today, London continues to attract Irish people seeking jobs in construction, media, finance, and health care.

Many Jamaicans immigrated to the United Kingdom in the late 1940s after a hurricane devastated the country and its economy. The British government invited Jamaican immigrants to relieve a labor shortage in British cities. In more recent decades, many affluent Jamaicans have moved to the United Kingdom to study or to work as professionals. However, others continue to come to escape poverty and violence and to work difficult, low-paying jobs.

Immigration Push-Pull Factors

IMMIGRANTS	PUSH FACTORS	PULL FACTORS
Indian		
Bangladeshi		
Irish		
Jamaican		

1. **Synthesize** What in British history links these four countries? How does this help explain why people from these countries are among London's largest immigrant populations?

_____

2. **Draw Conclusions** Based on the passage, what can you conclude about the strength of the United Kingdom's economy since World War II?

_____

© NGSP & HB

SECTION **2** GOVERNMENT & ECONOMICS

2.1–2.4 Review and Assessment

Use with Europe Today, Sections 2.1–2.4, *in your textbook.*

Follow the instructions below to review what you have learned in this section.

Vocabulary Next to each vocabulary word, write the letter of the correct definition.

1. _____ tariff
2. _____ sovereignty
3. _____ exchange
4. _____ democratization
5. _____ privatization
6. _____ demographic
7. _____ aging population
8. _____ assimilate

A. the characteristic of a human population

B. the process of becoming a government in which the people hold political authority

C. a tax paid on imports and exports

D. the process of transferring control of government-owned businesses to private owners

E. to be absorbed into a society's culture

F. a trend that occurs as the average age of a population rises

G. to convert money from one currency to another

H. a state's control over its own affairs, free from outside control

Main Ideas Use what you have learned about Europe's government and economics to answer these questions.

9. **Explain** What did European countries discover after forming the Organization for European Economic Cooperation?

_____

_____

10. **Compare** How does the European Union's economy compare to others around the world?

_____

_____

11. **Identify** What are the three branches of the European Union's government?

_____

_____

12. **Analyze Cause and Effect** How has the adoption of the euro affected tourism within Europe?

_____

_____

13. **Describe** What did Greece and Ireland agree to do in 2010 to help manage their debts?

_____

_____

14. **Identify** Which power controlled much of Eastern Europe after World War II?

_____

_____

15. **Synthesize** What is the result of Europe's improved medical care and smaller families?

_____

_____

16. **Analyze Cause and Effect** How did the fall of communism in Eastern Europe affect migration within Europe?

_____

_____

© NGSP & HB

Focus Skill: Make Inferences

Answer the questions below to make inferences about the costs and benefits of European unification.

17. Why would the formation of the Common Market have encouraged trade across borders?

_____

_____

18. Why is having a stable democracy that respects human rights a requirement for joining the European Union?

_____

_____

19. Why do you think many Eastern European countries want to adopt the euro?

_____

_____

20. Why do you think Poland's nonviolent methods in the 1981 rebellion against its Communist government were successful?

_____

_____

21. What reasons might there be to explain the difficulties Yugoslavia encountered in its transition to democracy?

_____

_____

22. Why might many older Eastern Europeans disagree with younger people about their preference between communism and democracy?

_____

_____

23. How has Europe's aging population led to increased immigration?

_____

_____

24. What types of businesses in particular would need additional workers because of Europe's aging population?

_____

_____

Synthesize: Answer the Essential Question

What are the costs and benefits of European unification? Recall what you have read about political and economic agreements among European countries. Be sure to consider the transition across Eastern Europe from communism to democracy, as well as the uneven economic results that have taken effect.

_____

_____

_____

© NGSP & HB

SECTION **2** GOVERNMENT & ECONOMICS

2.1–2.4 **Standardized Test Practice**

Use with Europe Today, Sections 2.1–2.4, in your textbook.

Follow the instructions below to practice test-taking on what you've learned from this section.

Multiple Choice Circle the best answer for each question from the options available.

1. Which nation was a founding member of the European Union?

   A the Netherlands
   B Turkey
   C Ukraine
   D Norway

2. What year was the euro launched?

   A 1957
   B 1992
   C 1999
   D 2011

3. Why has Norway decided not to join the European Union?

   A It did not want to use the euro.
   B It did not want to lose sovereignty.
   C It is not part of Europe.
   D It has an unstable government.

4. Which countries make up the eurozone?

   A the countries that adopted the euro as their currency
   B the countries that signed the Treaty of Maastricht
   C the original members of the Common Market
   D the countries that have joined the EU but do not use the euro

5. By how much has trade increased since 2002 as a result of switching to the euro?

   A 2 percent
   B 10 percent
   C 25 percent
   D 40 percent

6. Which of the following best describes life in Eastern Europe under Soviet control?

   A Citizens enjoyed great democratic freedoms and high standards of living.
   B Citizens enjoyed great democratic freedoms but had low standards of living.
   C Citizens lacked democratic freedoms but had high standards of living.
   D Citizens lacked democratic freedoms and had low standards of living.

7. Which event occurred in 1991?

   A The Ukrainian people staged the Orange Revolution.
   B The Soviet Union collapsed.
   C Yugoslavia broke up into several countries.
   D Poland rebelled against its Communist government.

8. What is NATO?

   A a military alliance among European and North American countries
   B an economic partnership among European countries
   C a branch of the United Nations that promotes scientific and cultural cooperation
   D a nongovernmental organization dedicated to world health

9. Europe's aging population is an example of the continent's changing _____.

   A sovereignty
   B democratization
   C privatization
   D demographics

10. What is the goal of Australia's skilled migration program?

   A to eliminate barriers to international trade
   B to find work for Australians in Europe
   C to fill gaps in the country's workforce
   D to decrease the average age of the population

© NGSP & HB

Document-Based Questions

This excerpt from the article **Europe Is Shifting to a Migration Destination**, by Thomas Hayden, appeared in *National Geographic*'s 2010 issue of *EarthPulse*. Read the excerpt and answer the questions that follow.

> *In 2007, immigration accounted for 80 percent of Europe's population growth, helping to bridge the age gap as population growth declined in countries such as Spain, Portugal, Italy, Germany, and the Scandinavian nations. Younger immigrants in Europe typically pay more in taxes than they require in government services, helping to support social security payments to aging citizens.*

Constructed Response Read each question carefully and write your answer in the space provided.

11. In 2007, was immigration sufficient to make up for the decreasing population sizes in Spain and Portugal? Why or why not?

_____

12. According to the passage, how do native Europeans benefit from immigration?

_____

Extended Response Read each question carefully and write your answer in the space provided.

13. The passage focuses on the period before the global economic crisis of 2008 began. As jobs became scarce during the recession, what effect do you think this had on the relationships between immigrants and native Europeans?

_____

_____

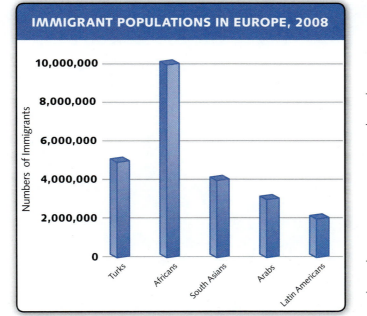

Source: UN Statistics, 2008

14. What was the combined number of immigrants from Turkey and Arab countries living in Europe in 2006? How does this combined number compare with the number of immigrants from Africa?

_____

_____

15. Why do you think people from Africa and Turkey make up the largest immigrant populations in Europe? How does this compare to the immigrant populations in North America that you have read about?

_____

_____

Name Class Date

Europe Geography & History

QUIZ: SECTION 1 GEOGRAPHY

Multiple Choice Circle the best answer for each question from the choices available.

1 Why is Europe called a peninsula?
A It is surrounded on three sides by water.
B It is the main landmass of Eurasia.
C It has an unusually long coastline.
D It has many mountains and rivers.

2 What generalization can be made about Europe's climate?
A Warm winds and ocean currents keep the entire continent very mild.
B Mistral winds keep most of the continent cold and dry.
C Currents, winds, and latitude result in varied climates.
D The winters everywhere are long and cold.

3 Why has Europe conducted so much trade and exploration?
A The climate is mild in all regions.
B Most places have easy access to water.
C People build dikes and create polders.
D Religious beliefs make trade important.

4 Why are the Danube and Rhine rivers important?
A They can be used easily for transportation.
B They run along mountain chains.
C They serve as borders between countries.
D They support agriculture and wildlife.

5 Why did fishing industries develop in Europe?
A The climate was too cold for farming.
B Europeans lacked grazing land for cattle.
C The market for seafood was excellent.
D Europe had plenty of ocean coastlines.

6 What has harmed the natural environment of the Mediterranean Sea?
A large fish damaging the red coral
B pollution, overfishing, overdevelopment
C the creation of marine reserves
D exploration, conservation, transportation

Constructed Response Write the answer to each question in the space provided.

7 How did the people of the Netherlands increase their farming industry?

____

____

8 Why do so many people live on the Northern European Plain?

____

____

BLACK-AND-WHITE COPY READY

Name _____ Class _____ Date _____

Europe Geography & History
QUIZ: SECTION ❷ EARLY HISTORY

Multiple Choice Circle the best answer for each question from the choices available.

1 Why were the Greek city-states independent from one another?
  A Cleisthenes made them all democracies.
  B They had different religions and education systems.
  C The tyrants who ruled them hated each other.
  D Geography made traveling and communicating difficult.

2 What happened during the golden age of Greece?
  A Democracy, wealth, and culture flourished in Athens.
  B Alexander the Great conquered Greece.
  C Sparta became the center of wealth and the arts.
  D Tyrants ruled all the city-states.

3 When Rome was a republic, what were its branches of government?
  A patricians, plebeians, and tribunes
  B the Senate and the Assembly
  C executive, legislative, and judicial
  D the Twelve Tables

4 Which fact indicates that the Roman Empire was the most powerful empire in the ancient world?
  A It was peaceful during Augustus' rule.
  B Its rule extended over three continents.
  C It made Christianity the official religion.
  D It defeated the barbarians from Germany.

5 What were the most powerful influences on Europeans during the Middle Ages?
  A Johannes Gutenberg and Martin Luther
  B Charlemagne and the Franks
  C Roman Catholic Church and feudalism
  D Eastern Orthodox Church and the plague

6 What was "reborn" during the Renaissance?
  A arts, philosophy, culture
  B war between Eastern and Western Europe
  C Europeans' hatred of tyrants
  D feudalism, manorialism, and the Crusades

Constructed Response Write the answer to each question in the space provided.

7 What helped bring about the end of the feudal system?

_____

_____

8 What were the results of Martin Luther's objections to some corrupt Catholic practices?

_____

_____

© NGSP & HB

Europe Geography & History

QUIZ: SECTION ❸ EMERGING EUROPE

Part 1: Multiple Choice Circle the best answer for each question from the choices available.

1 What did Prince Henry of Portugal start?
A a public education system
B a school of navigation
C the factory system
D the Columbian Exchange

2 What happened during the Industrial Revolution?
A Farmers produced many more crops.
B Poor people overthrew the rich and took over government.
C Canals were dug to connect waterways.
D Workers produced goods by machine in factories.

3 Who started the French Revolution?
A the First Estate, priests
B the Second Estate, nobles
C the Third Estate, working people
D Napoleon Bonaparte

4 What kinds of rights are described in the Declaration of Independence and the Declaration of the Rights of Man and of the Citizen?
A women's right to vote and to hold office
B natural rights such as life, liberty, and property
C equal rights for men, women, and all races
D workers' rights such as safety and equal pay

5 Which country was blamed for World War I and punished by the Treaty of Versailles?
A Austria
B Bulgaria
C Germany
D Italy

6 Which countries were America's enemies during World War II?
A Germany, Italy, Japan
B Poland, Austria, France
C Great Britain, France, Russia
D Ireland, Spain, Sweden

Constructed Response Write the answer to each question in the space provided.

7 What were good and bad results of the Industrial Revolution?

_____

_____

8 How was the Cold War different from World War II?

_____

_____

© NGSP & HB

Europe Geography & History
CHAPTER TEST A

Part 1: Multiple Choice Circle the best answer for each question from the choices available.

1   Which phrase best describes the
    Mediterranean climate?
    A polar temperatures, short summer
    B hot dry summers, mild rainy winters
    C long cold winters, cool summers
    D heavy rains, high humidity

2   What body of water has lost much of its red
    coral because of overfishing, overdevelopment,
    and pollution?
    A Atlantic Ocean
    B Baltic Sea
    C Mediterranean Sea
    D North Sea

3   What allowed Philip of Macedonia
    to conquer Greece?
    A The Golden Age made Greeks lazy.
    B A long war between Sparta and Athens
    weakened Greece.
    C The gods turned against the Greeks.
    D Volcanoes erupted, killing much of the
    Greek population.

4   How was the Roman Republic different
    from earlier governments in Italy?
    A Tyrants made people obey or die.
    B Citizens had rights and roles in government.
    C An emperor ruled through force.
    D Women ran different branches of
    government.

5   What was the purpose of the Crusades?
    A to punish German farmers
    B to expand the Roman Empire
    C to convert Franks to Christianity
    D to take back holy lands from Muslims

6   What industry in Great Britain started
    the Industrial Revolution?
    A textile
    B coal mining
    C fishing
    D iron working

7   What was one of the main purposes of
    European exploration?
    A to gain knowledge from other lands
    B to trade and establish new colonies
    C to start cities in warmer climates
    D to fight wars with other countries

8   What was an important cause
    of the French Revolution?
    A rivalry with other countries
    B religious groups that became very radical
    C unhappy workers in factories
    D unfair treatment of lower and middle
    classes

9   What was one result of the treaty
    that ended World War I?
    A Several new countries were formed.
    B Germany declared victory and gained
    wealth.
    C Great Britain won colonies in India.
    D Russia became part of the Soviet Union.

10 What was the Holocaust in World War II?
    A the fall of the Berlin Wall
    B a battle between Russians and Germans
    C the bombing of Hiroshima, Japan
    D the killing of 6 million Jews and others

Europe Geography & History

CHAPTER TEST A

Part 2: Interpret Maps Use the map and your knowledge of Europe to answer the questions below.

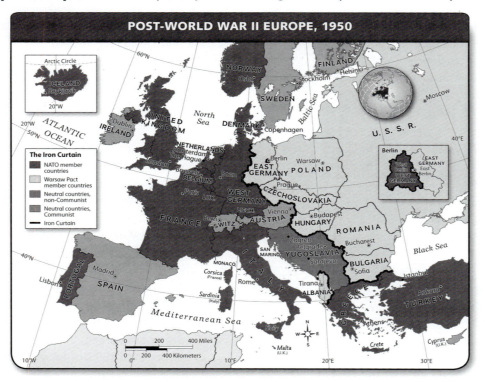

11 To what group did the countries shaded darkest belong?
 A the Warsaw Pact
 B NATO
 C neutral Communist countries
 D neutral non-Communist countries

12 East Berlin belonged to what group?
 A NATO
 B the Warsaw Pact
 C neutral Communist countries
 D neutral non-Communist countries

13 In the inset map, what does the line around West Berlin show?
 A It is inside East Germany.
 B It was part of the Warsaw Pact.
 C It was surrounded by the Iron Curtain.
 D It was a neutral non-Communist country.

Constructed Response Use a complete sentence to write the answer in the space provided.

14 Look at the Warsaw Pact countries. What generalization can you make about their physical position in Europe?

_____

_____

Name _____ Class _____ Date _____

Europe Geography & History

CHAPTER TEST A

Part 3: Interpret Charts Use the chart and your knowledge of Europe to answer the questions below.

MILES OF RAILWAY TRACK IN SELECTED EUROPEAN COUNTRIES (1840–1880)			
	1840	1860	1880
Austria-Hungary	144	4,543	18,507
Belgium	334	1,730	4,112
France	496	9,167	23,089
Germany	469	11,089	33,838
Great Britain	2,390	14,603	25,060
Italy	20	2,404	9,290
Netherlands	17	335	1,846
Spain	0	1,917	7,490

Source: Mordern History Sourcebook

15 Which country had the most extensive transportation system in 1840?
A France
B Germany
C Great Britain
D Netherlands

16 In what year did Germany have more miles of track than Great Britain?
A 1820
B 1840
C 1860
D 1880

17 Which country had the biggest increase in miles of track between 1840 and 1880?
A Belgium
B Germany
C Great Britain
D Spain

Constructed Response Use a complete sentence to write the answer in the space provided.

18 Judging from this chart, which country had the most industry in 1880? Explain your answer.

_____

_____

© NGSP & HB

Europe Geography & History

CHAPTER TEST A

Part 4: Document-Based Question Use the documents and your knowledge of Europe during World War II to answer the questions below.

Introduction

World War II took place from 1939–1945. More people, both soldiers and civilians, died in this war than in any other war—between 50 million and 70 million. It was waged all over the world, in Europe, Asia, and the Pacific Islands. Winston Churchill became prime minister of Great Britain early in the war.

Objective: Describe the human cost of World War II.

DOCUMENT 1 Quotation from Winston Churchill's first speech as prime minister to the British House of Commons, May 13, 1940

> You ask, what is our policy? I can say: It is to wage war, by sea, land and air, with all our might . . . . You ask, what is our aim? I can answer in one word: It is victory, victory at all costs, victory in spite of all terror, victory however long and hard the road may be; for without victory, there is no survival.

Source: http://www.winstonchurchill.org/learn/speeches/speeches-of Winston-

Constructed Response Use complete sentences to write the answer in the space provided.

19 What does Churchill say will happen if Britain does not achieve victory?

_____

_____

20 What words tell you that Churchill thinks the war will be very difficult?

_____

_____

DOCUMENT 2 Quotation from *The Face of War*, by Martha Gellhorn, reporter during World War II

> The mountains of Italy are horrible . . . . The jeep driver . . . has to collect the wounded from the sides of these mountains, carry them down on stretchers and drive them back to base hospitals over roads which would be dangerous even if no one shelled them. In these mountains, it is not unusual for a wounded man to be carried for ten hours before he reaches a road and a waiting ambulance.

Source: *The Face of War*, Martha Gellhorn, 1988, Atlantic Monthly Press

Constructed Response Write the answer to each question in the space provided.

21 How did Italy's geography further endanger wounded soldiers?

_____

22 What dangers could occur during the jeep drive down the mountain?

_____

DOCUMENT 3 Military Casualties, World War II

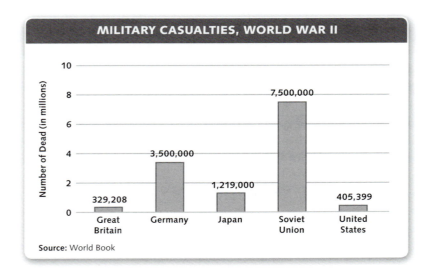

Source: World Book

Constructed Response Write the answer to each question in the space provided.

23 How many Soviet soldiers were killed during World War II?

_____

24 Japan and Germany lost the war. How did their combined losses compare to the combined losses of the victorious countries?

_____

Extended Response Write a paragraph to answer the question. Use information from all three documents and your knowledge of Europe in writing your paragraph. Use the back of this page or a separate piece of paper to write your answer.

25 Judging from the other documents, did Churchill's warning about the war come true? Explain.

Europe Geography & History

CHAPTER TEST B

Part 1: Multiple Choice Circle the best answer for each question from the choices available.

1 What do Great Britain, Iceland, and Corsica have in common?
- **A** Their climate is similar.
- **B** They are all islands of Europe.
- **C** Their people share a language.
- **D** They are all on plateaus.

2 What natural resources do the mountain chains of Europe provide?
- **A** canals and waterways
- **B** tropical plants and fish
- **C** forests and minerals
- **D** grains and cattle

3 What were the two most important city-states in Greece?
- **A** Sparta and Athens
- **B** Corinth and Delhi
- **C** Thebes and Eleusis
- **D** Ithaca and Olympia

4 Why was Alexander the Great famous?
- **A** wrote poetry and drama
- **B** improved the beauty and power of Athens
- **C** made democracy stronger
- **D** extended his empire through most of Eurasia

5 Who made the laws when Rome was a republic?
- **A** eight judges on the court
- **B** the Senate and the Assembly
- **C** two consuls in the executive branch
- **D** Tarquin the tyrant

6 What was Martin Luther trying to reform?
- **A** the Roman Catholic Church
- **B** the teachings of Islam
- **C** farms and fisheries
- **D** art and literature

7 What was the guillotine used for?
- **A** a military weapon
- **B** an industrial machine
- **C** an execution method
- **D** a farming tool

8 Why did cities grow so quickly during the Industrial Revolution?
- **A** People moved to cities for factory jobs.
- **B** Artists could make more money in the cities.
- **C** Droughts dried up good farmland.
- **D** Shipping businesses were failing.

9 What happened just before World War I?
- **A** Countries made alliances to increase their power.
- **B** The Industrial Revolution spread through Europe.
- **C** Europe was divided into East and West.
- **D** Poor people tried to overthrow the wealthy.

10 What event helped Adolf Hitler gain power in Germany?
- **A** the Cold War
- **B** the French Revolution
- **C** the Great Depression
- **D** the Russian Revolution

Europe Geography & History

CHAPTER TEST B

Part 2: Interpret Maps Use the map and your knowledge of Europe to answer the questions below.

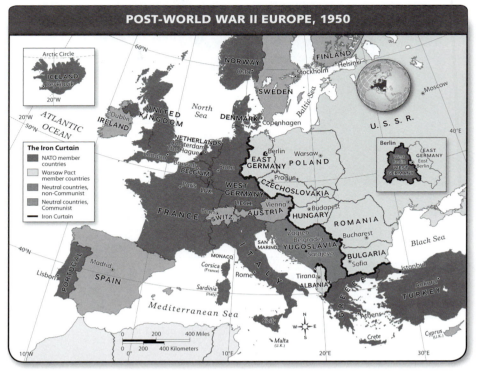

POST-WORLD WAR II EUROPE, 1950

The Iron Curtain
- NATO member countries
- Warsaw Pact member countries
- Neutral countries, non-Communist
- Neutral countries, Communist
- Iron Curtain

MAP TIP: Read the key, and notice the inset map of East and West Berlin.

11 East Berlin lay inside what country?
  A Czechoslovakia
  B East Germany
  C Poland
  D West Germany

12 What were the countries with no shading members of?
  A NATO
  B neutral Communist countries
  C neutral non-Communist countries
  D Warsaw Pact

13 How many Communist and non-Communist countries were neutral?
  A 3
  B 5
  C 7
  D 9

14 Which Warsaw Pact country was separated from the others by a non-member country?
  A Albania
  B Bulgaria
  C Czechoslovakia
  D Yugoslavia

© NGSP & HB

Europe Geography & History
CHAPTER TEST B

Part 3: Interpret Charts Use the chart and your knowledge of Europe to answer the questions below.

MILES OF RAILWAY TRACK IN SELECTED EUROPEAN COUNTRIES (1840–1880)			
	1840	**1860**	**1880**
Austria-Hungary	144	4,543	18,507
Belgium	334	1,730	4,112
France	496	9,167	23,089
Germany	469	11,089	33,838
Great Britain	2,390	14,603	25,060
Italy	20	2,404	9,290
Netherlands	17	335	1,846
Spain	0	1,917	7,490

Source: Mordern History Sourcebook

CHART TIP: Notice the years listed in the headings.

15 Which country had the most railway track in 1860?
- **A** Austria-Hungary
- **B** France
- **C** Germany
- **D** Great Britain

16 Which country had the least amount of railway track in 1840?
- **A** France
- **B** Great Britain
- **C** Italy
- **D** Spain

17 In 1880, how many more miles of railway track did Germany have than Great Britain?
- **A** 877
- **B** 8,778
- **C** 9,849
- **D** 18,778

18 Which country built the least amount of railway track during the period shown?
- **A** Belgium
- **B** France
- **C** Italy
- **D** Netherlands

© NGSP & HB

Name _____   Class _____   Date _____

Europe Geography & History

CHAPTER TEST B

Part 4: Document-Based Question Use the documents and your knowledge of Europe
to answer the questions below.

Introduction

World War II took place from 1939–1945. More people, both soldiers and civilians, died in this war than in
any other war—between 50 million and 70 million. It was waged all over the world, in Europe, Asia, and the
Pacific Islands. Winston Churchill became prime minister of Great Britain early in the war.

Objective: Describe the human cost of World War II.

DOCUMENT 1 Quotation from Winston Churchill's first speech as prime minister
to the British House of Commons, May 13, 1940

> You ask, what is our policy? I can say: It is to wage war, by sea, land and air, with all our might . . . .
> You ask, what is our aim? I can answer in one word: It is victory, victory at all costs, victory in spite
> of all terror, victory however long and hard the road may be; for without victory, there is no survival.

Source: http://www.winstonchurchill.org/learn/speeches/speeches-of Winston-

Constructed Response Write the answer to each question in the space provided.
You do not need to write complete sentences.

19 In what ways is Churchill calling the British to fight?

_____

_____

20 What is Churchill's goal, and why is that his goal?

_____

_____

© NGSP & HB

DOCUMENT 2 Quotation from *The Face of War,* by Martha Gellhorn, reporter during World War II

The mountains of Italy are horrible . . . . The jeep driver . . . has to collect the wounded from the sides of these mountains, carry them down on stretchers and drive them back to base hospitals over roads which would be dangerous even if no one shelled them. In these mountains it is not unusual for a wounded man to be carried for ten hours before he reaches a road and a waiting ambulance.

Source: *The Face of War,* Martha Gellhorn, 1988, Atlantic Monthly Press

Constructed Response Write the answer to each question in the space provided.

21 What steps were taken to carry a wounded soldier from the mountains of Italy to the hospital?

_____

22 How would the journey affect a wounded soldier?

_____

DOCUMENT 3 Military Casualties, World War II

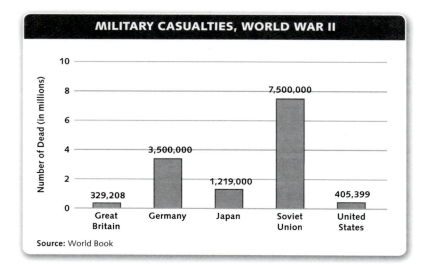

Source: World Book

Constructed Response Write the answer to each question in the space provided.

23 How many more Soviet soldiers were killed than German soldiers?

_____

24 The United States was in the war for a shorter time than Great Britain. How do American military losses compare with those of the British?

_____

Extended Response Write a paragraph to answer the question. Use information from all three documents and your knowledge of Europe in writing your paragraph. Use the back of this page or a separate piece of paper to write your answer.

25 In what ways did the war affect all the countries that fought, both the winners and losers?

BLACK-AND-WHITE COPY READY

Name Class Date

Europe Today

QUIZ: SECTION 1 CULTURE

Multiple Choice Circle the best answer for each question from the choices available.

1 The language group that includes Italian, French, and Spanish is called
A Germanic.
B Indic.
C Romance.
D Slavic.

2 The most widely practiced religion in Europe is
A Christianity.
B Hinduism.
C Islam.
D Judaism.

3 European art and music developed out of the artistic achievements of ancient
A France and Germany.
B Greece and Rome.
C India and Turkey.
D Japan and China.

4 The period of European art in which emotions were conveyed through landscapes and natural scenes was called the
A Abstract period.
B Baroque.
C Renaissance.
D Romantic period.

5 The number of books available to people increased due to the development in the 1400s of
A the modern novel.
B epic poetry.
C the printing press.
D writing paper.

6 The cuisine of any European country is largely determined by its
A landforms and climate.
B seaports and villages.
C economy and railroads.
D imports and exports.

Constructed Response Write the answers in the space provided.

7 What factor has contributed to the growth of Islam in Europe today?

____

____

8 How was European literature of the 20th century affected by World War I and World War II?

____

____

Europe Today

QUIZ: SECTION ❷ GOVERNMENT & ECONOMICS

Multiple Choice Circle the best answer for each question from the choices available.

1 Why was the European Union formed?
- **A** to treat countries as one nation
- **B** to unite Europe economically
- **C** to control how farmers get paid
- **D** to give workers equal wages

2 What is one of the benefits of using a single currency in many EU countries?
- **A** It allows people, money, and goods to move freely from country to country.
- **B** It helps prevent member countries from going into heavy debt.
- **C** It helps create an open market in which goods can be traded using a tariff.
- **D** It allows countries to maintain control of their own affairs.

3 What major event occurred in Eastern Europe in 1991?
- **A** the privatization of Romanian business
- **B** the collapse of the Soviet Union
- **C** the Orange Revolution in Ukraine
- **D** the rebellion of Polish workers

4 In what former country did a long civil war take place among ethnic groups?
- **A** Czechoslovakia
- **B** Hungary
- **C** Russia
- **D** Yugoslavia

5 Which of the following is an example of Europe's changing demographics?
- **A** Many of its countries have rebuilt their economies.
- **B** It has an aging population.
- **C** Many of its immigrants send money to relatives.
- **D** It has exports and imports.

6 Which of the following is one reason many people migrate to Europe?
- **A** to escape conflicts in their countries of origin
- **B** to purchase homes of their own
- **C** to have safe drinking water for their families
- **D** to sell goods from back home

Constructed Response Write the answers in the space provided.

7 Though some eastern European countries have had success changing to a market economy, what problems have other countries in the region faced?

_____

_____

8 What are three of Europe's immigration challenges?

_____

_____

© NGSP & HB

Europe Today

CHAPTER TEST A

Part 1: Multiple Choice Circle the best answer for each question from the choices available.

1 Which of the following statements about European cities is true?
   A Many share a common language.
   B Most are located along a coast.
   C Many began hundreds of years ago.
   D Most reflect a single cultural tradition.

2 During what era did a good deal of music and art focus on religion?
   A Middle Ages
   B Classical period
   C Romantic period
   D impressionism

3 Why was the development of the printing press so important?
   A It made people enjoy novels.
   B It provided jobs for workers.
   C It made books available to more people.
   D It led to other inventions.

4 In what location do olives, tomatoes, and garlic play a key role in cooking traditions?
   A Mediterranean countries
   B Great Britain
   C Scandinavian countries
   D Eastern Europe

5 Which of the following is a requirement for membership in the European Union?
   A The country must use the euro.
   B The country must be controlled by a powerful government body.
   C The country must have a stable democracy that respects human rights.
   D The country must charge fair tariffs.

6 Which statement about the EU is true?
   A It has a government with executive, legislative, and judicial branches.
   B It is made up of northern European countries.
   C It has immigration quotas that limit the populations of its members.
   D It is also known as the Common Market.

7 In 2010, what event showed cooperation among eurozone countries?
   A Citizens of countries with their own currency were allowed to travel in the eurozone.
   B Germany and France educated citizens on how to exchange their currencies.
   C Eurozone countries loaned money to Greece and Ireland to help them manage their debt.
   D Eurozone countries removed all tariffs and transaction fees for businesses.

8 Which of the following countries built strong democratic governments after their Communist dictatorship ended?
   A Croatia and Serbia
   B Poland and Hungary
   C Turkey and Greece
   D Ukraine and Bulgaria

9 What is one effect of Europe's aging population?
   A Workers are needed to replace senior citizens who are retiring.
   B There are fewer drivers than there were ten years ago.
   C Workers must work longer hours to support elderly parents.
   D There are more retirement homes than single-family homes.

10 How did the fall of communism affect European migration?
   A People started new businesses instead of moving to another country.
   B People from Great Britain's former colonies moved to England.
   C People had to develop new skills before they could move to another country.
   D People from Eastern Europe moved to Western Europe.

Europe Today

CHAPTER TEST A

Part 2: Interpret Maps Use the map and your knowledge of Europe to answer the questions below.

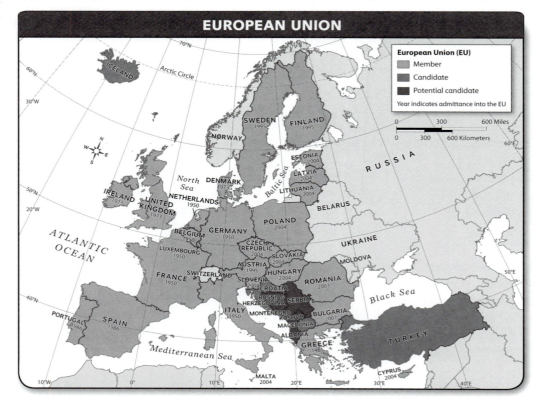

11 Which of the following nonmember countries is completely surrounded by EU members?

A Macedonia

B Moldova

C Switzerland

D Ukraine

12 Which of these countries became an EU member most recently?

A Bulgaria

B Finland

C Hungary

D Poland

13 In which of these countries would French tourists need to exchange their euros for another currency in order to buy something?

A Germany

B Greece

C Ireland

D Norway

Constructed Response Use a complete sentence to write the answer in the space provided.

14 How does Turkey differ geographically from other candidate countries?

_____

_____

Europe Today

CHAPTER TEST A

Part 3: Interpret Charts Use the chart and your knowledge of Europe to answer the questions below.

COST OF A TEN-MINUTE PHONE CALL TO THE U.S. IN EUROS (€)*		
Country	1997	2006
Belgium	7.50	1.98
Czech Republic	3.09	2.02
Denmark	7.41	2.38
Ireland	4.61	1.91
Spain	6.17	1.53
France	6.78	2.32
United Kingdom	3.50	2.23

* 1997 prices have been converted to euros
Sources: Eurostat

15 Between 1997 and 2006, what country experienced the greatest change in the price of a phone call to the United States?
  A Belgium
  B Denmark
  C France
  D Spain

16 In which two countries did the price of a phone call change the least between 1997 and 2006?
  A Czech Republic and the United Kingdom
  B Denmark and Czech Republic
  C France and the United Kingdom
  D the United Kingdom and Spain

17 In 2006, in which country would a person spend about 3 euros to make a 20-minute phone call?
  A Belgium
  B France
  C Ireland
  D Spain

Constructed Response Use a complete sentence to write the answer in the space provided.

18 What generalization can be made about the prices of phone calls in 2006 as compared with the prices in 1997?

_____

_____

© NGSP & HB

Europe Today

CHAPTER TEST A

Part 4: Document-Based Question Use the documents and your knowledge of Europe to answer the questions below.

Introduction

Europe today benefits in many ways from its culturally diverse population. However, this diversity also presents challenges. One group of people, called the Roma, migrated to Europe long ago yet still face serious problems in their adopted countries.

Objective: Analyze the challenges facing the Roma in Europe today.

DOCUMENT 1 Quotation about the Roma

With a population of 8 to 12 million, Roma are Europe's largest minority. But unlike other minority groups, such as Turkish workers in Germany, . . . the Roma have no nation to call home or to argue on their behalf. The Roma migrated out of India beginning around A.D. 1100. . . . By the 1300s, Roma were settled across Europe. Today, Roma are found in every European country.

Source: "Europe's Largest Minority Gaining Recognition, Expert Says," by John Roach. *National Geographic News* (nationalgeographic.com/news), January 4, 2007.

Constructed Response Use complete sentences to write the answers in the space provided.

19 Based on this passage, what inference can you make about the Roma today?

_____

_____

20 What is the author's attitude toward the Roma?

_____

_____

DOCUMENT 2 Quotation about the Roma

Roma are the victims of prejudice, often violent, at home in Eastern Europe. Thousands have migrated westward to seek a better life, particularly as the expansion of the European Union has allowed them to take advantage of freedom-of-movement rules. Yet, . . . most Roma are still worse off than under communism. . . . Today, conditions in Roma settlements on the edges of towns and villages [in Western Europe] rival Africa or India for their deprivation [poverty].

Source: "Europe's Roma: Hard Traveling." *The Economist* (economist.com), September 2, 2010.

Constructed Response Use complete sentences to write the answers in the space provided.

21 Why have many Roma left Eastern Europe?

_____

22 What can you conclude about the effect of Europe's political changes on the Roma?

_____

DOCUMENT 3 Chart of Roma Populations in Selected European Countries, 2010

COUNTRY	ROMA POPULATION	PERCENTAGE OF TOTAL POPULATION
Spain	725,000	1.57
France	400,000	0.62
Britain	265,000	0.43
Italy	145,000	0.24
Germany	105,000	0.13

Sources: Council of Europe, 2010

Constructed Response Use complete sentences to write the answers in the space provided.

23 How does the Roma population in France compare to the Roma population in Germany?

_____

24 To which country shown on the chart have most of the Roma migrated?

_____

Extended Response Write a brief paragraph to answer the question. Use information from all three documents and your knowledge of Europe in writing your paragraph. Use the back of this page or a separate piece of paper to write your answer.

25 What changes might improve life for the Roma in Europe?

Europe Today

CHAPTER TEST B

Part 1: Multiple Choice Circle the best answer for each question from the choices available.

1 Which of these has become the fastest-growing religion in Europe?
A Christianity
B Hinduism
C Islam
D Judaism

2 What did the technique of perspective allow artists to do?
A emphasize form and color over realism
B convey light and shadow
C portray the human figure more correctly
D give their work greater depth

3 Which form of music tells a story through words and music?
A concerto
B opera
C sonata
D symphony

4 Which early writer is famous for plays that are still performed today?
A Miguel de Cervantes
B Homer
C William Shakespeare
D Dante

5 Why is fish an important part of the cuisine of certain European countries?
A The countries are located near bodies of water where fishing takes place.
B Fish contains vitamins and minerals.
C The countries import different types of fish from around the world.
D Fish is inexpensive.

6 Which currency is used in at least 17 countries in Europe?
A euro
B franc
C mark
D pound

7 What is one advantage when countries share the same currency?
A Visitors and business people can travel without paying airport taxes or luggage fees.
B Countries can raise taxes without hurting their citizens.
C Visitors and business people can go from country to country using the same money.
D Countries can give up control of their own economies.

8 Which statement is true about eastern European countries today?
A Many want to join the European Union.
B Their state-run businesses continue.
C Their unemployment rates are very low.
D Many have returned to communism.

9 What is one cause of Europe's aging population?
A Older citizens are living with their children.
B People are retiring earlier.
C Governments encourage immigration of older people.
D People are living longer.

10 Which word describes what happens when people from one culture are absorbed into another society's culture?
A assimilate
B demographic
C migrate
D privatization

Europe Today

CHAPTER TEST B

Part 2: Interpret Maps Use the map and your knowledge of Europe to answer the questions below.

MAP TIP: Members and candidates are shaded on the map. The darkest shading represents potential candidates, countries that would like to join the EU but are not yet being considered.

11 Which of the following countries are potential candidates for the EU?
  A Austria and Greece
  B Belarus and Ukraine
  C Iceland and Turkey
  D Serbia and Albania

12 In what year did both Ireland and the United Kingdom become EU members?
  A 1950
  B 1973
  C 1986
  D 2004

13 Where are most nonmember countries located?
  A Eastern Europe
  B Northern Europe
  C Southern Europe
  D Western Europe

14 The Arctic Circle passes through which two EU members?
  A Belgium and Germany
  B Finland and Sweden
  C Iceland and Greenland
  D Norway and Russia

© NGSP & HB

Europe Today
CHAPTER TEST B

Part 3: Interpret Charts Use the chart and your knowledge of Europe to answer the questions below.

COST OF A TEN-MINUTE PHONE CALL TO THE U.S. IN EUROS (€)*		
Country	1997	2006
Belgium	7.50	1.98
Czech Republic	3.09	2.02
Denmark	7.41	2.38
Ireland	4.61	1.91
Spain	6.17	1.53
France	6.78	2.32
United Kingdom	3.50	2.23

* 1997 prices have been converted to euros
Sources: Eurostat

CHART TIP: Remember that the prices listed are in euros, not U.S. dollars.

15 In 1997, in which country was a phone call to the United States the most expensive?
A Belgium
B Denmark
C France
D Spain

16 In 1997, in which country was a phone call to the United States the least expensive?
A Czech Republic
B Denmark
C Ireland
D the United Kingdom

17 In 2006, in which countries did a phone call cost less than 2 euros?
A the United Kingdom and Denmark
B Belgium, Ireland, and Spain
C Denmark, Belgium, and France
D Czech Republic and Belgium

18 In 2006, in which country would a person have paid the least for a phone call?
A Belgium
B France
C Ireland
D Spain

© NGSP & HB

Europe Today

CHAPTER TEST B

Part 4: Document-Based Question Use the documents and your knowledge of Europe to answer the questions below.

Introduction

Many people in Europe today originally came from other countries and cultures. This has created a rich and diverse European society, but sometimes problems have arisen. One group of people, known as the Roma, have faced difficulties that are sometimes overlooked.

Objective: Explain the challenges facing the Roma in Europe today.

DOCUMENT 1   Quotation about the Roma

With a population of 8 to 12 million, Roma are Europe's largest minority. But unlike other minority groups, such as Turkish workers in Germany, . . . the Roma have no nation to call home or to argue on their behalf. The Roma migrated out of India beginning around A.D. 1100. . . . By the 1300s, Roma were settled across Europe. Today, Roma are found in every European country.

Source: "Europe's Largest Minority Gaining Recognition, Expert Says," by John Roach. *National Geographic News* (nationalgeographic.com/news), January 4, 2007.

Constructed Response Write the answer to each question in the space provided. You do not need to write complete sentences.

19 For how long have the Roma been settled in Europe?

_____

_____

20 How do the Roma differ from other minority groups in Europe?

_____

_____

DOCUMENT 2 Quotation about the Roma

Roma are the victims of prejudice, often violent, at home in Eastern Europe. Thousands have migrated westward to seek a better life, particularly as the expansion of the European Union has allowed them to take advantage of freedom-of-movement rules. Yet, . . . most Roma are still worse off than under communism. . . . Today, conditions in Roma settlements on the edges of towns and villages [in Western Europe] rival Africa or India for their deprivation [poverty].

Source: "Europe's Roma: Hard Traveling." *The Economist* (economist.com), September 2, 2010.

Constructed Response Write the answer to each question in the space provided.

21 Based on this passage, has migration to Western Europe improved the lives of Roma who have settled there?

_____

22 Living conditions for the Roma in Western Europe are compared with living conditions in what other parts of the world?

_____

DOCUMENT 3 Chart of Roma Populations in Selected European Countries, 2010

COUNTRY	ROMA POPULATION	PERCENTAGE OF TOTAL POPULATION
Spain	725,000	1.57
France	400,000	0.62
Britain	265,000	0.43
Italy	145,000	0.24
Germany	105,000	0.13

Sources: Council of Europe, 2010

Constructed Response Write the answer to each question in the space provided.

23 Which country listed in the chart has the largest population of Roma?

_____

24 Based on the percentages shown, what generalization can you make about the total Roma population in each country?

_____

Extended Response Write a paragraph to answer the question. Use information from all three documents and your knowledge of Europe in writing your paragraph. Use the back of this page or a separate piece of paper to write your answer.

25 How would you summarize the situation facing the Roma of Europe?

© NGSP & HB

GEOGRAPHY & HISTORY

SECTION 1.1 PHYSICAL GEOGRAPHY

Reading and Note-Taking

1. Humid Temperate; Dry Summer; Humid Cold; Tundra & Ice; Unclassified Highlands
2. Student has drawn a reasonably accurate outline map of Europe and has accurately shaded the map, using a different shade for each climate region.
3. Western Uplands, Northern European Plain, Central Uplands, and Alpine region are accurately labeled on the map.
4. Italian, Scandinavian, and Iberian peninsulas are correctly labeled on the map.
5. Atlantic Ocean, Mediterranean Sea, Black Sea, and North Sea are accurately labeled on the map.

Vocabulary Practice

Suggested rubric:

1. Student accurately describes Europe's physical geography.
2. Student correctly defines peninsula as a piece of land that is surrounded on three sides by water.
3. Student correctly defines uplands as an area of high land made up of hills, mountains, and plateaus.
4. Student explains that Europe is known as a "peninsula of peninsulas."
5. Student explains that the Western Uplands stretch from the Scandinavian Peninsula to Spain and Portugal.
6. Student explains that the Central Uplands are located at the center of Europe.
7. Student may also include details about climate, such as describing sirocco winds and the mistral; a description of Mediterranean climate or the colder climate of Eastern Europe, Greenland, and northern Scandinavia.

GeoActivity Map Europe's Land Regions

1. Students' maps should resemble the following:

2. Students' responses should show an understanding of the relationship between what people can produce based on the landforms of a region. For example, the economy of the Western Uplands would likely be more industrial because hydroelectric power is produced there.

SECTION 1.2 A LONG COASTLINE

Reading and Note-Taking

Trade: Having so much water access helped the growth of trade, which was central to the growth of the civilizations of ancient Greece and Rome. Their early sailors brought back goods and ideas from other lands that greatly influenced European culture.

Industry: Europe developed industries, including fishing and farming, which depend on oceans and seas.

Exploration: Europe's location near oceans and seas encouraged exploration, which helped Europe's rulers build empires, obtain raw materials, and spread their religious beliefs.

Settlement: People settled around coastal ports to be close to centers of trade and industry.

Vocabulary Practice

Word: bay
Definition: a small body of water set off from the main body of water
In Your Own Words: a body of water that is smaller than a gulf
Sentence: A bay is surrounded on three sides by land and is a good place for a port to dock ships and for people to settle and make a living by fishing.

Word: fjord
Definition: a narrow bay set between cliffs, steep hills, or mountains
In Your Own Words: a finger of water between high land
Sentence: Norway has fjords, which are narrow bays.

Word: polder
Definition: land that has been reclaimed, or taken, from the seabed
In Your Own Words: a dried seabed that is used for farming in the Netherlands
Sentence: In the Netherlands, people figured out how to create polders to make farmland.

GeoActivity Analyze Early European Trade

1. Students' maps should resemble the following:

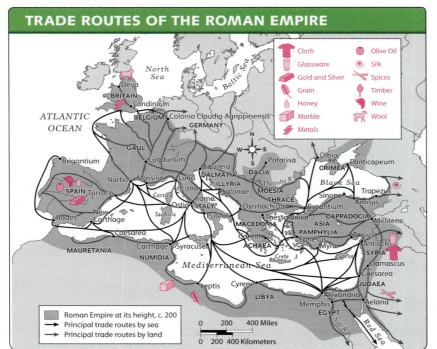

TRADE ROUTES OF THE ROMAN EMPIRE

2. Students' responses will vary based on the good they chose. They should be able to choose a variety of trade routes, both by land and by sea. They should also correctly name major cities along the routes.
3. Location positively affects trade when good roads can be built and there is access to sea routes.

SECTION 1.3 MOUNTAINS, RIVERS, AND PLAINS

Reading and Note-Taking

K: Europe has a variety of landforms and natural resources.

W: What kinds of economic activities are supported by these landforms and natural resources?
L: Mountain chains provide natural resources such as forests and mineral resources. The valleys between the mountains in the chain provide fertile land for growing olives and other crops.

Europe's many rivers provide important transportation routes for travel, shipping, and irrigation. The fertile soil of the Northern European Plain is ideal for farming. The Plain also has some of the largest urban centers in Europe. These places attract tourists, and tourism would help the economy.

Vocabulary Practice

ACADEMIC VOCABULARY
Possible response: The river has become so narrow that it is barely navigable.

Venn Diagram

canal: a long, narrow waterway created by people for passage through an area of land
waterway: a river or canal wide enough and deep enough for shipping and boating
both: Both are water passages that are navigable; a canal is an example of a waterway.

GeoActivity Explore Amsterdam's Canals

1. flooding; they built a dam to control the water
2. The lack of dry land and the threat of flooding posed many problems. The complex series of canals made it easier for Dutch merchants to import goods from all over the world and export their own.
3. The canals are laid out in a pattern of semicircles. City planners might have used this pattern as the city expanded outward to add more space for shipping without changing what already existed.

SECTION 1.4 PROTECTING THE MEDITERRANEAN

Reading and Note-Taking

Problem: Human activities have harmed the Mediterranean Sea's natural environment.

Event 1: Overfishing
Event 2: Pollution
Event 3: Overdevelopment

Solution: Marine reserves like the Scandola Natural Reserve are protected areas where people are prohibited from fishing, swimming, or anchoring boats. This enables marine life to thrive and helps restore marine habitats.

Make a Prediction: If the Mediterranean Sea had more marine reserves, more of its marine habitats would recover from pollution, overfishing, and overdevelopment. People would complain about not being able to catch, sell, or eat as much fish as they used to. They might also find it hard to find an affordable place to live and work if development on the coast is decreased. Without affordable housing and work, more and more people would be forced to live elsewhere, perhaps away from the coasts.

Vocabulary Practice

ACADEMIC VOCABULARY

Possible response: Large numbers of people living along the Mediterranean coastlines has caused erosion.

Comparison Chart

ecosystem
Definition: a community of living organisms and their natural environment

marine reserve
Definition: an area of the sea where animals and plants are given special protection

Similarities: Both are particular areas of the environment that provide habitats for living things. A marine reserve is usually created to protect a part of the ocean's ecosystems.

GeoActivity Graph Fishery Catches in the Mediterranean

1. Students' graphs should resemble the following:

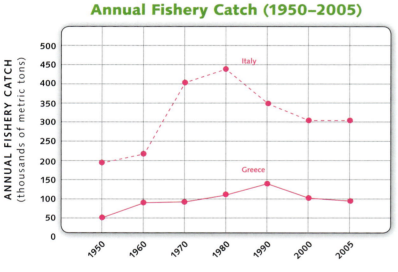

Annual Fishery Catch (1950–2005)

2. Fishery catches increased for both countries although catches for Greece were much lower than for those for Italy.

3. Catches decreased for both countries. Students may suggest the reasons mentioned in Section 1.4 in their textbook and in the passage: reduced fish populations because of overfishing, pollution, and development.

SECTION 1 REVIEW AND ASSESSMENT

Vocabulary

1. D	**3.** G	**5.** A	**7.** C
2. F	**4.** E	**6.** B	

Main Idea

8. the Western Uplands, Northern European Plain, Central Uplands, and Alpine region

9. Ocean currents bring warm temperatures, winds bring wet weather, and cold winds from France bring cold, dry weather.

10. A Mediterranean climate has mild, rainy winters. Eastern Europe has long, cold winters. Northern European areas have polar climates.

11. Water access promoted trade, which influenced the continent's culture with goods and ideas.

12. Possible response: European villages developed near bays because they were good locations for harbors, which promoted trade.

13. The mountain chains provide forests for wood, mineral resources, and fertile land.

14. The flat land of the plain provides a suitable area for living and for building cities.

15. Pollution and overfishing have negatively impacted the Mediterranean's ecosystem.

Focus Skill: Evaluate

16. Possible response: Europe's many peninsulas provided people with great access to water, which was important for trade and exploration. Water access gave rise to fishing and to farming.

17. Sailors brought back goods and ideas from other lands.

18. Polders increased the amount of usable land for farming and possibly for buildings. There would have been more agricultural products for trade and more places to build industries.

19. Possible response: Explorers used the water access to help build empires, obtain raw materials, and spread religious beliefs.

20. The mountain chains provide natural resources, such as wood and iron ore, which can be used for trade. The rich soil in the valleys between the mountains provides many crops that can be sold.

21. Canals enable easier transport between places and more trade, boosting economic activity.

22. Shipping goods on rivers and canals is probably faster and more economical than moving them over mountains or on roads from town to town.

23. Possible response: A growing population has led to overfishing, the erosion of Mediterranean coastlines, and the pollution of its waters.

Synthesize: Answer the Essential Question

Because Europe is a "peninsula of peninsulas," it provided people with great access to water. The water provided a means for sailors and explorers to visit other lands and interact with the people of other regions. Inland waterways, such as the Danube and Rhine rivers, made it possible to travel to other countries within the continent. But people also improved on their geographic situations, such as by building polders to create farmland in the Netherlands or building canals to connect rivers and make additional navigable waterways for trade.

SECTION 1 STANDARDIZED TEST PRACTICE

Multiple Choice

1. B	**3.** B	**5.** B	**7.** C	**9.** B
2. D	**4.** B	**6.** A	**8.** B	**10.** C

Constructed Response

11. As bluefin tuna is becoming scarce, its price is rising, and it is becoming more in demand as a delicacy.

12. Sylvia Earle thinks people would be willing to pay anything for the last remaining bluefin tuna.

Extended Response

13. Possible response: Bluefin tuna may be becoming scarce because of human activities. Reducing these activities and creating marine reserves would help bluefin tuna survive.

Data-Based Questions

14. According to the chart, most cities show marked increases in population by 2015, Istanbul in particular.

15. Possible response: According to the chart, populations are on the rise. To preserve ecosystems and long-term food sources, people must reduce pollution, overdevelopment, and overfishing.

SECTION 2.1 ROOTS OF DEMOCRACY

Reading and Note-Taking

oligarchy: a government ruled by a small group

monarchy: a government ruled by a monarch, such as a king or queen

tyranny: rule by a single person who has complete control of the government

mixed government: a government that practices two or more types of rule, such as limited democracy and oligarchy

direct democracy: a government in which all citizens vote directly for laws

limited democracy: a democracy in which only a certain group, such as adult males, are considered citizens with the right to vote.

assembly: a government body that makes laws

Vocabulary Practice

ACADEMIC VOCABULARY
Possible response: Aristocrats in ancient Greece sometimes set up oligarchies in the form of small ruling councils.

Definition and Details

Word: city-state
Definition: an independent community that has its own government and consists of a city and the area around it
Detail: Around 800 B.C., several Greek city-states started to thrive.
Detail: Greek city-states adopted different forms of government, including limited democracy, oligarchy, monarchy, tyranny, and mixed.

Word: democracy
Definition: a form of government in which people have the power to make decisions and vote for representatives
Detail: Athens, Eleusis, and Olympia practiced limited democracy.
Detail: Cleisthenes established a direct democracy, which allowed all citizens to vote directly for laws.

GeoActivity Analyze Primary Sources: Democracy

1. favoring the many instead of the few
2. that they are all created equal
3. Both put an emphasis on equality for all.
4. Works by Greek writers and philosophers might have helped persuade the founders that democracies can really work. They also might have convinced them of the importance of governments.

SECTION 2.2 CLASSICAL GREECE

Reading and Note-Taking

Democracy
Q: What did Pericles do to strengthen democracy?
A: He paid citizens to hold public office because as long as officials were unpaid, only the wealthy could afford to serve.

Architecture
Q: What is the Parthenon and who built it?
A: The Parthenon is a temple that Pericles built to honor the goddess Athena, who was believed to protect Athens.

Philosophy
Q: Who were Socrates and Plato?
A: They were the leading philosophers during the golden age of Greece.

Science
Q: What was Hippocrates' contribution to the practice of medicine?
A: Hippocrates changed the practice of medicine by insisting that illnesses originated in the human body and were not caused by evil spirits.

Vocabulary Practice

Word: golden age
Definition: a period of prosperity and cultural achievement
Sentence: Greece's golden age ended around 431 B.C. because the Peloponnesian War broke out between Athens and Sparta and weakened both city-states.
Example: Classical Greece
Synonym: heyday

Word: philosopher
Definition: a person who closely examines basic questions about the universe
Sentence: The philosopher tried to figure out the meaning of human existence.
Examples: Socrates, Plato
Synonym: thinker

GeoActivity Research Ancient Greek Contributions

Students' research results will vary but may include some of the following examples:

- Art and architecture: architecture of public buildings, columns, marble or bronze statues showing ideal beauty, painted vases

- Government: direct democracy, elections, self-rule, voting, concept of citizenship, public assemblies, Plato's *Republic*
- Health: gymnastics, emphasis on sports, science in medicine, Hippocratic Oath
- Literature and drama: plays, comedies, tragedies, poetry, written history, mythology, *Aesop's Fables*
- Science and philosophy: mathematics, geometry (Euclid, Pythagoras), physics, astronomy, Socrates, Plato, Aristotle, "Socratic method"
- Sports: Olympic Games, specific sports such as discus-throwing, wrestling, gymnastics

SECTION 2.3 THE REPUBLIC OF ROME

Reading and Note-Taking

Topic: The Roman Republic Government

Main Idea: The Roman Republic created a form of government that Europe and the West would later follow.

Executive: Two consuls elected for one-year term by Assembly after 490 B.C.

Legislative: At first only patricians, wealthy landowners, could take part in the government. They made laws and controlled the Senate, which was made up of 300 patricians, who were selected by the consuls to serve as their advisors for life. In 490 B.C., plebeians gained the right to form an assembly and elect legislative representatives called tribunes. The assembly also had the right to make laws and select consuls.

Judicial: Eight judges, each serving for one year, oversaw the lower courts and governed the provinces.

Legal Code: Around 450 B.C., the government published the Twelve Tables. These bronze tablets set forth the rights and responsibilities of Roman citizens

Citizens: At the time the Twelve Tables were published, only adult male landowners born in Rome were citizens. Roman women had a limited form of citizenship and could not vote or hold public office.

Vocabulary Practice

ACADEMIC VOCABULARY

In the Roman republic, one consul could veto decisions made by the other consul.

Compare/Contrast Paragraph

Students' paragraphs will vary but should use all three vocabulary words correctly. They should explain that a republic is a form of government in which citizens elect officials who govern according to law. Students should accurately compare and contrast the roles, duties, and rights of patricians and plebeians in the Roman Republic, noting that these were two classes of people: patricians were wealthy landowners; plebeians were mostly farmers.

GeoActivity Compare Greek and Roman Governments

1. **Similarities:** Both are democratic—citizens have a say in their government.

Differences: A direct democracy is a system of government in which every citizen who is eligible to vote can vote on laws. It is the rule of the majority. In a republic, an elected representative votes on behalf of the citizens.

2. The Assembly allowed poorer citizens to have a say in laws. It ensured that the interests of the wealthy were balanced by the interests of those who were not as wealthy.

3. The United States was large in size. It was easier and more realistic for people to elect representatives to go to the capital than to go themselves.

4. In a direct democracy, sometimes the majority will look out for its interests only, and smaller groups outside the majority have no rights. In a republic, the representative may not know what the people want or may be swayed to vote in a way that does not accurately represent citizens' views.

SECTION 2.4 THE ROMAN EMPIRE

Reading and Note-Taking

Creation of the Empire
44 B.C.: Members of the Senate stab Caesar to death and civil war begins; Caesar's nephew Octavian fights in the war, which puts an end to the Roman Republic

Rise of the Empire
27 B.C.: Octavian becomes Emperor Augustus; Pax Romana, the "Roman peace," begins

Decline of the Empire
Around A.D. 235: Rome has a series of poor rulers; barbarians begin invading from the north
330: Emperor Constantine moves capital of weakened empire from Rome to Byzantium, which he renames Constantinople. Constantine makes Christianity lawful throughout the empire.
395: The empire is divided into an Eastern Empire and a Western Empire with two different emperors
476: Invaders overthrow the last Roman emperor and end the Western Empire.

Rome left a rich legacy of technology and language. Modern engineers and architects still use Roman technology such as the arch to construct buildings today, and many of the roads in the network that connected the Roman Empire are still in use. Modern Romance languages such as Spanish and Italian are based on Latin, the language spoken in the Roman Empire. Many English words have Latin roots.

Vocabulary Practice

W: aqueduct
D: a structure that looks like a bridge and that is used to carry water
S: Roman engineers built aqueducts to carry water across the empire.
W: barbarian
D: a member of an ancient German tribe
S: Barbarians from the north began to invade the Roman Empire.

GeoActivity Analyze the Roots of Modern Languages

1. a. dis-; b. bon; c. ped; d. ped; e. sub; f. spect
2. a. to make laws; b. dom; domestic cat; c. the distance someone's feet travel; d. under the sea; e. against it
3. Students' responses will vary but may include the following words: perspective, retrospective, prejudice, judicious, legal, illegal, expedition, extrovert, vertigo, domesticate, dislike, submerge, and other words with dis- or sub- as prefixes.

SECTION 2.5 MIDDLE AGES AND CHRISTIANITY

Reading and Note-Taking

Subject: The Middle Ages

The Roman Catholic Church
- helped unite people during the Middle Ages
- played a leading role in government by collecting taxes
- made its own laws
- waged wars
- began a series of Crusades to take back lands in Southwest Asia from Muslim control

The Feudal System
- provided a social structure during the Middle Ages
- provided security for each kingdom
- was organized like a pyramid with the king at the top, followed by the lords who pledged allegiance to the king, and then the vassals who pledged loyalty and service to the lords
- Serfs, who farmed the lord's land in return for shelter and protection, lived on the lord's manor and were at the bottom of the feudal system pyramid.

The Growth of Towns
- helped end the feudal system
- Trade and business developed, and people began to leave the manors.
- The bubonic plague, which swept through Europe in 1347, killed millions and greatly reduced the workforce in the towns.
- Employers were desperate for workers and offered higher wages, which drew many farmers from the country to work in towns.

Vocabulary Practice

Feudal System: a social system in Europe in which people worked and fought for nobles who gave them protection and the use of land in return

1st tier: King: owned vast territory
2nd tier: Lords: powerful nobles who owned land, lived on estates called manors
3rd tier: Vassals: were given pieces of land by lords in exchange for their loyalty and service for the lord; Some vassals served as knights, who were warriors on horseback.
4th or bottom tier: Serfs: farmed the lord's land in return for shelter and protection; lived on the lord's manor; dwelt in small huts; had to give most of the crops they grew to the lord of the manor

GeoActivity Categorize Effects of the Crusades

1. **Economy:** trade in luxury goods; Italian cities grew wealthy; merchants got rich
   Society: weakening of feudal system; loss of nobles' wealth and influence; knowledge of new foods, goods, and ideas
   Technology: building in stone; better ships
   World View: awareness of other cultures and places
   Shifts in Power: king acquired nobles' estates; more power for Pope and Church
2. Possible response: It was beneficial because new ideas came to Europe as a result.

SECTION 2.6 RENAISSANCE AND REFORMATION

Reading and Note-Taking

Factors
1. Towns helped end the feudal system.
2. The Roman Catholic Church started to lose power.
3. Some merchants grew wealthy from increased trade in growing towns, allowing them to become patrons of the arts.
4. The Byzantine Empire fell, and scholars from the empire came to Italy, bringing with them ancient writings of the Greeks and Romans.
5. Studies of the ancient writings encouraged humanism.
6. Johannes Gutenberg developed a printing press that printed many books in a short amount of time and gave more people access to knowledge.

Effects
1. Growth of art: painters began to use perspective to make paintings look three-dimensional.
2. Growth of architecture: Architects used elements of ancient Greek and roman design.
3. Growth of literature: Writers wrote in the vernacular, or everyday language (not Latin.)
4. People began to look more critically at the Church.
5. Martin Luther objected to corruption in the Church and started the Reformation.

Vocabulary Practice

Word: indulgence
Definition: a payment relaxing the penalty for sin, sold by the Church
In Your Own Words: the Church's pardon in return for money
Sentence: Martin Luther was shocked at the Church's practice of selling indulgences.

Word: perspective
Definition: artistic technique that makes a painting look as if it has three dimensions
In Your Own Words: a way that makes the image in a flat painting or drawing look more realistic
Sentence: During the Renaissance, artists used perspective to make their paintings seem more life-like.

GeoActivity Map the Protestant Reformation

Students' maps should resemble the following:

MAJOR PROTESTANT RELIGIONS, c. 1650

Major Protestant Religions, c. 1650
- Anglicanism
- Calvinism
- Lutheranism

1. Lutheranism; Anglicanism
2. because Rome, the center of the Roman Catholic Church, was located in southern Europe

SECTION 2 REVIEW AND ASSESSMENT

Vocabulary

1. D   **3.** A   **5.** E   **7.** F
2. G   **4.** B   **6.** C

Main Ideas

8. Athens and Sparta
9. Pericles wanted to strengthen democracy, expand the empire, and beautify Athens. He paid citizens to hold public office, built a strong navy, and rebuilt the city.
10. Alexander the Great, who loved Greek culture, took over and expanded the empire, spreading Greek ideas.
11. Both the judicial and executive branches served one-year terms; the judicial branch governed the lower courts and the provinces, while the executive branch led the government.
12. Possible response: The Roman Way were values that helped the Roman people become fine, upstanding citizens.
13. the invasion of German tribes, or barbarians
14. People left the manors to work in the towns, which weakened the feudal system.
15. The printing press made the spread of ideas much easier, which helped the flowering of culture that occurred during the Renaissance.

Focus Skill: Draw Conclusions

16. The city-states were isolated by mountains and so they developed unique communities and governments.

17. Possible response: They followed and liked Pericles because he gave both rich and poor citizens an equal opportunity to hold office.
18. Possible response: Socrates and Plato continue to influence thought and philosophy today.
19. Possible response: Plebeians may not have gotten along with patricians because they could not take part in government.
20. Possible response: Julius Caesar was one of the first strong rulers. He helped the poor. His murder helped give rise to the Roman Empire.
21. Possible response: The Roman Empire built roads to create easier trade and transport of its army. It had ambitious rulers and a powerful military.
22. People could read or hear stories and information in the language they spoke. This would have been appealing and spread ideas to more people.
23. Possible response: People had the freedom to gain knowledge and had more exposure to art, architecture, and literature.

Synthesize: Answer the Essential Question

As empires rose and fell, Europe developed democracy, in which people had more of a voice in decisions about their lives. Later, people drew upon ancient ideas as they created governments. Cultural achievements and advances of the past also shaped Europe. Painting, literature, and technology (the printing press) helped people become more educated and expressive.

SECTION 2 STANDARDIZED TEST PRACTICE

Multiple Choice

1. A   **3.** B   **5.** A   **7.** C   **9.** B
2. D   **4.** C   **6.** C   **8.** D   **10.** A

Constructed Response

11. Landless serfs were bound to a life of hard work and little freedom.
12. A serf would no longer be the property of a noble.

Extended Response

13. Possible response: They were courageous enough to try a different life. They disagreed with the unequal social structure and were sick of their lives and lack of choice and freedom of movement.
14. executive, legislative, and judicial
15. Possible response: Two classes of people could take part in government; however, plebeians were restricted to the Assembly and patricians to the Senate.

SECTION 3.1 EXPLORATION AND COLONIZATION

Reading and Note-Taking

I. Purpose of Exploration
A. to find gold and establish trade with Asia
B. to convert native peoples to Christianity

II. Dangers of Exploration
A. Small, weak ships
B. Disease and attacks by native peoples in unknown lands
III. Explorers
A. Portuguese: Bartolomeu Dias, Vasco da Gama sailed along African coast
B. Italian: Christopher Columbus uncovered the continents of North America and South America
C. French: Jacques Cartier explored North America
D. English: Sir Francis Drake sailed around the world
IV. Colonies
A. Spanish: in Mexico and South America
B. French and English: North America
C. European countries also controlled parts of Africa and Asia

Vocabulary Practice

ACADEMIC VOCABULARY
Possible response: One purpose of the European voyages of exploration was to convert native peoples to Christianity.

Words in Context

1. The French and English set up colonies in North America.
2. Navigation is the skill of steering ships and reading maps to get to particular locations.
3. The Spanish established colonies in Mexico and South America.
4. Prince Henry's school of navigation marked the start of the European Age of Exploration because it taught sailors how to steer their ships and read maps.
5. By 1650, Europeans had established colonies in South America, North America, Africa, and Asia.

GeoActivity Compare European Explorers

1. **Cabot:** Italy; England; established British claims in Canada
   Balboa: Spain; Spain; claimed the Pacific Ocean and its shores for Spain
   Cabral: Portugal; Portugal; discovered and claimed Brazil for Portugal
   Vespucci: Italy; Spain and Portugal; traveled twice to South America
   Cartier: France; France; made claims in North America for France
2. They hoped to gain land and resources.
3. Possible response: Explorers needed to be daring and brave to endure hardships and long periods away from home. They also needed to be excellent sailors and navigators.
4. They were all long and probably dangerous. They were funded by a king or by a country because the ships, sailors, and materials needed would have been expensive.

....

SECTION 3.2 THE INDUSTRIAL REVOLUTION

Reading and Note-Taking

Pros: Standards of living rose; a prosperous middle class grew; new technologies and the factory system made production of textiles and other goods much more efficient and profitable.

Cons: Factory workers faced harsh conditions; laborers worked as many as 16 hours a day; child labor was common, and some child workers were chained to their machines; workers' living conditions were crowded and unsanitary.

Opinion: The changes of the Industrial Revolution were both positive and negative.

Vocabulary Practice

Suggested rubric for Cause and Effect Paragraph:

1. Student uses and correctly defines both Key Vocabulary words.
2. Student describes textile production before the Industrial Revolution as being largely home-based.
3. Student describes the impact of the stream-powered spinning wheel and spinning jenny.
4. Student describes the factory system of textile production after the Industrial Revolution, showing the effect of the Industrial Revolution on the textile industry.

GeoActivity Evaluate Industrial Revolution Inventions

1. Students' charts should include information on how each invention was used and its impact on society. Possible responses regarding the inventions' impact on society might include the following: improving the quality of life; making goods more cheaply; improving the speed of communication and the overall speed of manufacturing; the faster transport of goods; poor working conditions in factories; the loss of handcrafting jobs; and the increase in factory jobs.
2. Students' discussions will vary but should show that students are thinking beyond the initial implications of an invention.
3. Students' responses will vary. Possible response: The steam engine made the Industrial Revolution possible, but the telephone and assembly line had more of a lasting impact.

....

SECTION 3.3 THE FRENCH REVOLUTION

Reading and Note-Taking

Root 1: Harvests were poor and prices skyrocketed.
Effect 1: Mobs stormed the Bastille.
Effect 2: The Revolution started.

Root 2: The Third Estate paid most of the taxes, but had no voice in government.
Effect 1: The Third Estate demanded reforms.
Effect 2: King Louis the XVI refused to make reforms.

Root 3: The Enlightenment
Effect 1: the American Revolution
Effect 2: The Third Estate formed the National Assembly, which issued the Declarations of the Rights of Man and of the Citizen.

The Reign of Terror exhausted the French people and weakened the French government, giving Napoleon the opportunity to seize power.

Vocabulary Practice

Word: guillotine
I Read: The Jacobins used a machine called the guillotine to cut off the heads of an estimated 40,000 people.
I Know: The French executed Maximilien Robespierre, the Jacobin leader who led the Reign of Terror.
And So: Robespierre was probably executed by guillotine himself.

Word: radical
I Read: In 1792, the Jacobins, a group of radicals, or extremists, seized power and formed the National Convention.
I Know: Radicals favor extreme changes in government.
And So: The National Convention was much more extreme than the National Assembly.

GeoActivity Map Napoleon's Empire

1. Students' maps should resemble the following:

NAPOLEON'S EMPIRE, 1810

2. The battle and victory at Austerlitz show that Napoleon was able to move his troops deep into the territory of the powerful Austrian Empire.
3. Possible response: Great Britain is an island. To take any British territory, Napoleon would have had to attack by sea, which was not his strength.

SECTION 3.4 DECLARATIONS OF RIGHTS

Reading and Note-Taking

Document 1
Main Idea: All people are born equal, and they have guaranteed natural rights.
Detail: People are born with the right to life, liberty, and the pursuit of happiness.
Detail: Governments are formed to be sure the people have these rights.

Detail: Governments are given power by their citizens.

Document 2
Main Idea: The National Assembly of France declares that humankind has natural, guaranteed, and sacred rights.
Detail: People are born and remain free and equal.
Detail: Social differences should only promote the general good.
Detail: A government's purpose is to protect the human rights of liberty, property, security, and the resistance to oppression.

Vocabulary Practice

apartheid
What is it? a former social system in South Africa that denied black South Africans their rights
Why is it important? It went against the idea of unalienable human rights.
What is an example? laws that made blacks live apart from whites and that limited their movement

natural rights
What is it? rights that all people possess at birth
Why is it important? It ensures that people can move freely and not be enslaved.
What is an example? the right to move freely, be treated equally

GeoActivity Analyze Primary Sources: Women's Rights

1. Women have the same gift of reason as men do and should therefore be able to enjoy the same rights.
2. Women will revolt because they will not obey laws that they had no say in creating.
3. No, they did not. Adams states that if the new American government included rights for women, it would be kinder to women than the previous government, which was the colonial government set up by Britain.
4. The documents in the textbook focus on political freedom and economic equalities between men and governments but do not consider the inequality between men and women in society.

SECTION 3.5 NATIONALISM AND WORLD WAR I

Reading and Note-Taking

1. Nationalism sweeps through Europe in the 1800s.
2. Italian states form a unified Italy in 1870.
3. Led by Prussia, German states come together as a unified German empire in 1871.
4. Nationalism creates fierce competition between rival nations for raw materials and colonies.
5. Britain, France, and Russia form an alliance, the Triple Entente.
6. The German Empire and Austria-Hungary form an alliance, the Central Powers.
7. In 1914, Archduke Franz Ferdinand of Austria-Hungary is assassinated by a nationalist from Bosnia-Herzegovina who wanted to unite

with Serbia.

8. Austria-Hungary declares war on Serbia.
9. Russia, a Serbian ally, declares war on Austria-Hungary.

Vocabulary Practice

ACADEMIC VOCABULARY

Possible response: The neighborhood crime watch is an alliance in my community.

Definition Clues

nationalism
1. Nationalism is a strong sense of loyalty to one's country.
2. a feeling of loyalty and pride in your country
3. Possible response: Nationalism in the 1800s led both Italy and Germany to unification.
4. Not only was Archduke Franz Ferdinand assassinated by a nationalist, but nationalism led nations to compete for raw materials and colonies and to form alliances. So when one nation, such as Serbia, entered into a conflict, its allies felt that they had to join the fight.

trench
1. Both sides fought from trenches, or long ditches that protected soldiers from the enemy's gunfire.
2. a long ditch dug in the ground and used to shield soldiers from enemy sight and gunfire
3. The trenches in the battlefield seemed to go on for miles.
4. Trenches gave soldiers a place to take cover from the machine guns and tanks used by both sides.

GeoActivity Analyze Causes and Effects of World War I

Students' charts should contain the following causes and effects:

Intense nationalism → Alliance system → Triple Alliance, Triple Entente

Assassination of Archduke Ferdinand of Austria → Austria-Hungary declares war on Serbia → Russia declares war on Austria-Hungary

Communists gain control of Russia → Russia quits the war → United States joins the war

1. European nations were competing for colonies and resources, and tensions led them to take sides and form alliances.
2. Possible responses: No—Austria-Hungary would never have declared war on Serbia, which caused other countries in alliances to also declare war; Yes—tensions were so great that some other cause would have set off the war.
3. Germans were angry about the treaty, which caused more tension between Germany and the rest of Europe and led to World War II.

SECTION 3.6 WORLD WAR II AND THE COLD WAR

Reading and Note-Taking

Possible responses for Time Line:

1939: Germany invades Poland.
Great Britain and France declare war on Germany.
Germany conquers Poland.
Germany quickly defeats most of Europe, including France.

1941: Japan attacks the United States at Pearl Harbor.
The United States enters the war on the side of the Allies.

May 8, 1945: Germany surrenders.
Allied troops find concentration camps and other evidence of the Holocaust.
The United States drops atomic bombs on Hiroshima and Nagasaki in Japan.

September 2, 1945: Japan surrenders, ending the war.

Vocabulary Practice

Word: concentration camp
Definition: a type of prison where large numbers of civilians are kept during a war and are forced to live in very bad conditions
Word in Context: In the last months of the war, Allied troops were stunned to find the Nazi concentration camps where six million Jews and other victims had been murdered.
In Your Own Words: a prison camp holding non-soldiers during a war
Sentence: It is amazing that any prisoners survived the horrible conditions of the Nazi concentration camps.

Word: reparations
Definition: money that a country or group that loses a war pays because of the damage, injury, or deaths it has caused
Word in Context: As you have read, the Treaty of Versailles also placed full blame for the war on the country and forced it to pay reparations, or money to cover the losses suffered by the victors.
In Your Own Words: the losing side's payment for all the damage it caused during a war
Sentence: Reperations are one way to make the losing side pay for war.

GeoActivity Analyze the Results of World War II: Berlin

1. Using the map scale, students will likely estimate 20 miles dividing the city and an additional 40 miles surrounding West Berlin. In actuality, 28 miles divided the city and an additional 75 miles surrounded West Berlin.
2. West Berlin was like an island; it was surrounded by East Germany and Soviet influence.
3. The blockade of West Berlin and the Berlin Wall represented the tension that grew between the United States and the Soviet Union. They showed that the Soviet Union did not want a free exchange of people, ideas, or economic activity with the West.

SECTION 3 REVIEW AND ASSESSMENT

Vocabulary

1. D
2. E
3. B
4. H
5. F
6. C
7. G
8. A

Main Ideas
9. Possible response: gold, spices, silk, gems

10. Great Britain; new inventions and technologies

11. Possible response: Cities grew rapidly because of work and factories. There were more products. Standards of living rose for some, but at first living conditions were not good for workers and their families.

12. First Estate: priests; Second Estate: nobility; Third Estate: everyone else—merchants to peasants

13. Napoleon overthrew the French government after the violent years of the French Revolution. He made himself emperor and led the military against other European powers.

14. Possible response: Natural rights means that from birth people have the right to be who they want to be, live where they want to live, and be free.

15. the assassination of Archduke Franz Ferdinand of Austria-Hungary

16. Axis Powers: Germany, Italy, Japan; opponents: Allies (Great Britain, Soviet Union, United States)

Focus Skill: Make Inferences

17. Possible response: Europeans leaders wanted to extend their beliefs and increase their cultural influence.

18. Possible response: The land Columbus came upon may have been new to him, but it was not new to its native inhabitants.

19. Setting up a colony was a way to claim more land for one's country and to keep other countries from claiming it.

20. Producing cloth before the industrial revolution was time consuming. Everyone wears clothes so there would always be a need for cloth. With large machines, more cloth could be produced and sold more quickly.

21. Possible response: Although others may have liked and profited by making work faster and easier, factory workers worked in poor conditions and worked long hours for very low wages. They may have been grateful for the work but not the conditions they worked in.

22. Possible response: It is important for all people to be treated equally and to have the freedom to live their lives in ways that bring them happiness.

23. Germany, beaten and angry about having to pay reparations and lose territories, may have wanted revenge and to reclaim their power.

24. A class system provides stability in society, but inequality eventually can lead to rebellion and strife.

Synthesize: Answer the Essential Question

Europeans explored other lands and made an effort to colonize and convert their people. The innovations of the Industrial Revolution spread outside Europe, changing people's lives. Ideas of political thinkers and philosophers spread as well, often causing major political and social change. Finally, European nations engaged in wars to exert power, involving and affecting nations around the world.

SECTION 3 STANDARDIZED TEST PRACTICE

Multiple Choice

1. C	3. D	5. C	7. D	9. B
2. A	4. A	6. C	8. A	10. D

Constructed Response

11. According to the article, Marie Antoinette was touched by such a lovely welcoming.

12. The poor people are overwhelmed by taxes.

Extended Response

13. Possible response: Marie Antoinette, being of noble rank, surprisingly does not seem to look down on "the poor people" who are far below her rank but instead seems to understand their problems.

14. Machine technology allowed people to extract mineral ores and turn them into usable metal products.

15. Possible response: Belgium ranked second. Belgium may have had almost as many rich deposits of coal to power steam engines as Britain had.

FORMAL ASSESSMENT

SECTION 1 QUIZ

1. A	3. B	5. D
2. C	4. A	6. B

7. They built dikes to hold back the sea and created polders by draining water from land, which was then often used for farming.

8. It has fertile soil for growing crops and is quite large, so it contains many farms; large cities grew up where there were so many people.

SECTION 2 QUIZ

1. D	3. C	5. C
2. A	4. B	6. A

7. Towns grew; serfs and farmers left manors to find jobs in trades and businesses; the bubonic plague killed millions in towns, which reduced the workforce, so farmers went to cities to make higher wages.

8. Protestant churches were formed, and the Roman Catholic Church started a Counter-Reformation.

SECTION 3 QUIZ

1. B	3. C	5. C
2. D	4. B	6. A

7. Good: Cities grew, standards of living rose, and a prosperous middle class grew. Bad: Factory workers worked in bad conditions for too many hours a day, children worked in factories, neighborhoods were crowded, and open sewers spread diseases.

8. The two main enemies of the Cold War, the United States and Soviet Union, never fought each other directly; there were threats and political tensions. In World War II, enemy nations fought each other directly in large battles.

GEOGRAPHY & HISTORY TEST A

Part 1: Multiple Choice

1. B	**3.** B	**5.** D	**7.** B	**9.** A
2. C	**4.** B	**6.** A	**8.** D	**10.** D

Part 2: Interpret Maps

11. B   **12.** B   **13.** C

14. They lie in the eastern part of Europe and have poorer access to oceans; some are landlocked.

Part 3: Interpret Charts

15. C   **16.** D   **17.** B

18. Germany did, because it had the most miles of railway track. Industry was needed to manufacture track, and the railways were used to transport resources and products.

Part 4: Document-Based Question
Constructed Response

19. Britain will not survive.

20. "victory at all costs," "in spite of all terror," "however long and hard the road may be"

21. When they were injured in the mountains, it was a long, hard, and painful journey back to hospitals; many probably died along the way.

22. They could get shelled by the enemy.

23. 7,500,000

24. They had a smaller number of soldiers killed: fewer than five million as opposed to the Allies' loss of more than eight million.

Extended Response

25. Yes, Churchill's warning about the war came true. Although the Allies achieved victory, the cost in human lives and injuries was terrible. It was terror-filled, a long hard road. The number of military deaths is one way of showing how terrible the war was.

GEOGRAPHY & HISTORY TEST B

Part 1: Multiple Choice

1. B	**3.** A	**5.** B	**7.** C	**9.** A
2. C	**4.** D	**6.** A	**8.** A	**10.** C

Part 2: Interpret Maps

11. B   **12.** D   **13.** C   **14.** A

Part 3: Interpret Charts

15. D   **16.** D   **17.** B   **18.** D

Part 4: Document-Based Question
Constructed Response

19. by sea, land, and air, with all their might

20. victory, because Britain will not survive if it loses the war

21. A jeep driver picked up the soldier from the mountain, carried him down on a stretcher to the jeep, and drove back to the base hospital over a long journey.

22. He could be in great pain for a long time, or he could die.

23. 4 million

24. The United States lost more soldiers than did Great Britain (about 76,000 more).

Extended Response

25. The cost for all countries was huge. The human and property damage was terrible for everyone, winner or loser. The winning countries had many more military deaths than did Germany and Japan. This demonstrates that Churchill was correct in saying that the war would be a long and hard road.

TODAY

SECTION 1.1 LANGUAGES AND CULTURES

Reading and Note-Taking

Central Concept: Europe has great cultural diversity.

Supporting idea: Languages vary across the continent.
Detail: three main language groups (Romance, Germanic, Slavic)
Detail: Some countries have more than one language.
Detail: Some languages have several dialects.

Supporting idea: Europe has a variety of religions and celebrations.
Detail: 45% of the population is Catholic.
Detail: Islam is the fastest growing religion.

Supporting idea: Most Europeans live in cosmopolitan cities.
Detail: Over 95% of Belgians live in cities.
Detail: European cities are old and were built differently from those in the U.S.
Detail: Europeans use public transportation more than Americans do in the U.S.

Vocabulary Practice

ACADEMIC VOCABULARY
Possible sentence response: London's international population helps give the city a cosmopolitan feel.

Word Map

Word: dialect
What is it? a regional variety of a language
What is it like? local speech, same language but different pronunciations
How does it help you? Knowing that a language has a dialect can explain why it might be hard to translate or understand.

Word: heritage
What is it? the traditions or beliefs that are part of a country's or people's history
What is it like? what you inherit; what's passed down; past music or literature
How does it help you? Knowing a heritage helps one understand the customs that are passed from generation to generation.

GeoActivity Compare Urban Development

1. approximately 8.75 square miles (approximately 3.5 miles east-west by 2.5 miles north-south; both maps have the same scale)
2. **Similarities:** Both developed along water; both have public transportation; both have parks.
   Differences: Chicago's layout has more of a pattern. Its streets form a grid system. There also appears to be more land between streets in Chicago. London's streets are short and close together. They don't have any pattern.
3. Possible response: People who live in Chicago probably find it easier to drive in the city because of the street pattern; people in London might rely more on public transportation to avoid driving. This would

be supported by the larger number of rail stations in London.

SECTION 1.2 ART AND MUSIC

Reading and Note-Taking

European Art
1. Ancient Greek and Roman depictions of gods and goddesses inspired later artists.
2. Art in the Middle Ages focused on religious subjects.
3. Renaissance painters used perspective to create a sense of depth.
4. 1800s: Romantic artists focused on natural scenes. Impressionists of late 1800s used light and color to capture a moment.
5. Modern abstract artists of 1900s emphasized form and color over realism.

European Music
1. Ancient Greek and Roman musicians used a few simple instruments.
2. The violin was introduced during the Renaissance.
3. Opera began during the Baroque period (1600–1750) when music became more complex.
4. The Classical and Romantic periods followed the Baroque and lasted until around 1910.
5. Beethoven and other composers employed techniques still in use today.

Make Inferences: These artists and musicians probably responded to the same societal changes, and may have influenced one another.

Vocabulary Practice

1. These modern artists often worked in an abstract style, which emphasized form and color over realism.
2. pertaining to ideas or emotions, and not concrete reality
3. We could not identify a person or thing in the drawing, so it must have been abstract.
4. An abstract painting could have large areas of color or lines that don't take a recognizable shape. It would not use perspective because that would mean it would be trying to show something realistically as if it were in three dimensions.
5. In ancient Greece and Rome, musicians played on simple instruments and singers sang. In the Middle Ages, because music was about religion, it probably had music with words. Troubadours of this time influenced the music of the Renaissance. This period was followed by the Baroque, in which opera was born. Later periods included singing, because opera is still around today.

GeoActivity Recognize Architectural Movements

1. Modernist; the building has a simple appearance with very little ornamentation. It appears to have been built out of steel, glass, and concrete.
2. Gothic; the cathedral is extremely tall with many pointed arches and stained-glass windows.
3. Neoclassical; the structure has simplified forms with columns and a dome, like the styles from ancient Greece and Rome.

SECTION 1.3 EUROPE'S LITERARY HERITAGE

Reading and Note-Taking

Main Idea: European literature, including poetry, plays, and novels, has reflected new ways of thinking over the centuries.

Details:
- Ancient Greeks and Romans wrote epic poems about national or cultural heroes.
- Writers such as Shakespeare and Cervantes explored human behavior during the Renaissance.
- Enlightenment writers focused on the rights of the individual.
- Many 19th-century writers addressed social issues, such as poverty and gender roles.
- Modern writers in the 20th century experimented with new forms and examined the inner workings of the mind.

Vocabulary Practice

Possible response:

Compare/Contrast Paragraph: European literature began with Greek and Roman epic poems, which are long poems about the adventures of heroes or heroines important to a culture. Virgil, a Roman, created the *Aeneid*, which told of the founding of Rome. Dante, who lived much later, wrote *The Divine Comedy*, an epic poem about religious and political ideas of his time. Another genre of literature is the novel, which is also a long work. Novels have characters and a plot. Like an epic poem, a novel can also be about the adventures of heroes and heroines, but it is written in a different form. Novels can be about emotions, nature, or about stories besides adventures. They can include humor, as in the novel *Don Quixote*, or deal with serious social problems like poverty, as in *Oliver Twist*. Epic poems and novels have some things in common, but the novel has had a longer history, has a less formal structure than an epic poem, and therefore is more able to be customized to express different ideas.

GeoActivity Analyze Primary Sources: Romantic Writing

1. Keats suggests the poem should focus on the subject matter and should not draw too much attention to its form or style. The subject of the poem should startle and amaze the reader, not the poem itself.
2. Shelley agrees with Keats. She believes that a poet should be a mirror that reflects and magnifies the beauty of nature.
3. Possible response: Romantic paintings would be realistic. The painters would want the viewer to focus on the subject of the painting rather than the style or form of the painting.

SECTION 1.4 CUISINES OF EUROPE

Reading and Note-Taking

Mediterranean Countries
- climate good for growing olives, tomatoes, garlic
- France and Italy known for sauces

Germany and the British Isles
- meats and noodles with heavy gravies
- potatoes are common side dish

Scandinavia
- herring and other fish are common
- herding culture provides deer and elk meat

Russia
- climate allows for root vegetables to be grown
- borscht is traditional beet soup

Hungary
- fertile soil used to grow grains and potatoes
- variety of breads and dumplings eaten
- national dish is beef stew called goulash

Ukraine
- grows wheat and other grains
- highly decorated breads made for holidays

Vocabulary Practice

Word: cuisine
Definition: the traditional cooking style of a culture; Example: Mediterranean cuisine

Word: staple
Definition: a basic food that is part of people's diets; wheat, fish

Similarities: Both address traditional parts of people's diets; both may be important to a culture's identity.

GeoActivity Graph Olive Oil Production Rates

1. in the 2000s; demand for olive oil increased as people learned about its health benefits
2. Australia: 1,554 percent; Chile: 89 percent; Greece: 21 percent; Italy: 9 percent; Spain: 122 percent; United States: 147 percent
3. Students' graphs should resemble the following:

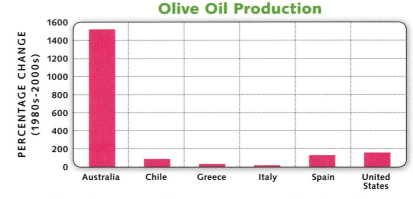

4. Australia; Italy; new-producing countries tended to have large percentages of increase; with the exception of Spain, old-producing countries had small percentages of increase
5. Possible response: The increased popularity will let old producers export their products to new markets and give them higher profits; however, the increase in new producers means that the old producers will have more competition, which could lower their profits.

SECTION 1 REVIEW AND ASSESSMENT

Vocabulary
1. E **3.** C **5.** F **7.** G
2. H **4.** A **6.** B **8.** D

Main Ideas
9. higher
10. Eastern Europe
11. religious subjects
12. Baroque
13. The printing press increased the popularity of books and their ideas.
14. They rejected traditional styles and experimented with new forms.
15. Cuisines in cooler regions tend to be heavier and more filling.
16. Seafood is probably more common in Portugal, located on the Atlantic coast, than in landlocked Czech Republic.

Focus Skill: Draw Conclusions
17. The large number of languages may help maintain a wide variety of distinct cultures.
18. Having multiple languages might divide the culture of a country such as Belgium.
19. Renaissance painters used perspective to create a more realistic sense of depth and space.
20. Baroque painting is likely highly detailed and precise.
21. Modern or 20th-century writers examined the unpredictability of life, the inner thoughts of characters, and experimented with new forms of poetry, plays, and novels. In the 19th century, writers focused more on social issues.
22. These works took a much more realistic look at life and focused on societal roles and social issues.
23. Local cultures and traditions clearly change over time and often include outside influences.
24. Ukraine has fertile soil and a climate that allows the country to grow great amounts of wheat and other grains.

Synthesize: Answer the Essential Question
Europe's diversity is reflected in its cultural achievements, as many of its major artistic, literary, and musical movements involved contributions from people across the continent who influenced one another. Europe's countries have maintained separate cultures, reflecting local geography and local customs, yet they share many influences. Europe's various cuisines also demonstrate the continent's rich diversity, yet as with language families, the similarities among some cuisines show the ways people across national borders share common roots.

SECTION 1 STANDARDIZED TEST PRACTICE

Multiple Choice
1. B **3.** C **5.** C **7.** A **9.** D
2. D **4.** D **6.** C **8.** B **10.** B

Constructed Response
11. The Mediterranean diet promotes social interaction, as shared meals are central to social customs and festivities.
12. Fruits and vegetables are a larger part of the Mediterranean diet than meat and dairy. The passage says that a moderate amount of fish, dairy, and meat characterize the cuisine, indicating that these items are not used in abundance.

Extended Response
13. The Mediterranean diet includes staple ingredients, such as olive oil, that are not found in the cuisines of countries with different climates. However, we can assume that the social importance of the cuisine and its importance to many customs are features that the Mediterranean diet shares with other cuisines across the continent.
14. Fish and seafood are healthiest, and should be eaten at least twice each week. [Red] Meat is the least healthy, and should only be eaten in small amounts on a monthly basis.
15. The Mediterranean diet is quite healthy, but may require people in different climate regions to import many ingredients from long distances away. This shipping may increase the prices of these ingredients and may have negative environmental consequences.

SECTION 2.1 THE EUROPEAN UNION

Reading and Note-Taking
I. **The European Union's Origins**
A. Organization for European Economic Cooperation forms in 1948
B. The Common Market: Western Europe sought closer economic ties
C. Treaty of Maastricht: Common Market countries form EU in 1992
II. **Features of the EU**
A. 27 member countries as of 2010
B. Government with three branches
C. Tariffs eliminated among most members
D. Euro: common currency used by 17 countries
III. **Membership**
A. Requirements: Having stable democracy that respects human rights
B. Some countries have declined joining to maintain sovereignty
C. Turkey has applied but is still being reviewed; Norway has chosen not to join to keep sovereignty.

Vocabulary Practice
ACADEMIC VOCABULARY
Possible sentence response: For the past few years, the euro has been a more valuable currency than either the U.S. or the Canadian dollar.

Words in Context
1. Countries may hesitate to trade with one another if the tariffs, or taxes on imports and exports, are too high.
2. Norway has not joined the EU due to concerns about sovereignty, as it does not want to lose control over its own affairs.
3. Sample sentence: The design of the different euros is very modern looking.
4. The country would lose some sovereignty, because it would have to

give up control over the value of its currency.

5. The Common Market eliminated tariffs on goods traded among member states.

6. Students' responses will vary.

GeoActivity Categorize Fundamental Rights

1. c. Equality
2. f. Justice
3. a. Dignity
4. e. Citizens' Rights
5. d. Solidarity
6. b. Freedoms
7. The governments of some countries might not appreciate being overruled by the Court of Justice. It also might be difficult for the court to investigate whether laws are being followed in each country.
8. Possible response: The EU is a large organization and could have a great influence on defending these rights worldwide, even in countries that do not defend them.

SECTION 2.2 THE IMPACT OF THE EURO

Reading and Note-Taking

1st Topic: Tourism; Tourists do not need to pay fees to exchange currency.; Outcome: Travel is easier and less expensive.

2nd Topic: Conducting business; Effect: Costs are easier to compare.; Outcome: Savings are passed along to consumers.

3rd Topic: Political effects; Effect: Greece and Ireland faced debt crises.; Outcome: Eurozone nations helped one another.

Conclusion: The euro has increased economic and political interdependence among eurozone countries.

Vocabulary Practice

ACADEMIC VOCABULARY
Sample sentence: Using the euro has made it easier for consumers to buy things in the eurozone.

Travel Blog

Students' blog entries will vary. Entries should demonstrate an understanding of the Key Vocabulary words and use each in an appropriate context. Students should indicate that Great Britain is not part of the eurozone, and that euros would need to be exchanged for British pounds when traveling there.

GeoActivity Graph the Economic Impact of the Euro

1. Students' graphs should resemble the following:

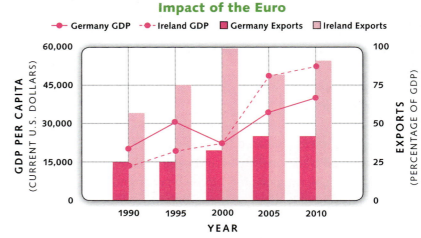

2. Trade was important because it made up the majority of GDP. It became almost the entire GDP after the adoption of the euro.

3. It may have helped people accept that the country would remain unified because it now shared a common currency with much of Western Europe.

4. Their economies should benefit greatly, as both Germany and Ireland experienced economic growth and increased trade after switching currencies.

SECTION 2.3 DEMOCRACY IN EASTERN EUROPE

Reading and Note-Taking

Government Transitions: After rebellions against the Communist government, countries declared independence.; Poland, Hungary, and the Czech Republic developed stable democracies.; Yugoslavia broke apart after civil war into several countries, including Serbia and Croatia.; Ukraine had an Orange Revolution and a leadership change but it did not hold. Voters put the former leader, Yanukovych, back into power.

Economic Transitions: Eastern European countries rebuilt their economies, changing from communist government-controlled economies to market economies. Government-controlled businesses became privately owned, through privatization. Not all countries had success. Poland has; it has a fast-growing economy. Some countries want to join the EU and NATO, but people in the countries do not always agree on the direction their countries should take.

Summary: Eastern Europe's path to democracy has been uneven, with some countries enjoying a much smoother transition than others. Government transitions to stable democratic governments have been the case for Poland, Hungary, and the Czech Republic, but Yugoslavia and Ukraine still experience challenges of civil war or unstable leadership. Not all countries of Eastern Europe have had success with privatization and creating competitive market-driven economies. Generally, some of the older generation believe that their countries were more secure under communist control. Others, generally of the younger generation, lean toward democracy.

Vocabulary Practice

Word: democratization
Definition: the process of becoming a government in which the people hold political authority
In Your Own Words: establishing a government in which people are fairly represented and leaders are held accountable
Use in Sentence: After the fall of communism, the process of democratization was apparently much smoother in Poland than in Ukraine.

Word: privatization
Definition: the process of transferring control of government-owned businesses to private owners
In Your Own Words: individuals or private companies assuming ownership of formerly public assets
Use in Sentence: Through privatization, independent companies gained control over the government-run automobile industry.

GeoActivity Research Reform Movements

Students' Project Organizers will vary based on the country they chose to research. Make sure students have gone through all the steps to complete the Project Organizer.

SECTION 2.4 CHANGING DEMOGRAPHICS

Reading and Note-Taking

1. Europeans are having smaller families.
2. The average age of the population is increasing.
3. Businesses need new workers to replace retirees.
4. Immigrant populations increase as people come to fill jobs.
5. Questions arise about assimilation and multiculturalism.
6. Many believe Europe should place limits on immigration.

Synthesize: Europe's changing demographics have created social stresses when immigrants enter a country and compete for jobs, and create a burden for governments helping to provide housing and health care for so many people. There are clashes among cultures, and people assimilate into the dominant culture slowly. The benefits of changing demographics have included bringing in people to fill jobs, exposing people to new cultural traditions, and building tolerance and cooperation among different cultures.

Vocabulary Practice

assimilate: to be absorbed into a society's culture
Possible sentence linking assimilate with demographics: Changing demographics raise questions about whether immigrants should assimilate into European cultures.

demographics: the characteristics of a human population, such as age, income, and education
Possible sentence linking demographics with aging population: Europe's changing demographics are reflected in its aging population.

aging population: a trend that occurs as the average age of a population rises

Possible sentence linking aging population with assimilate: As members of Europe's aging population retire, jobs become available for immigrants who must balance assimilating into European cultures and retaining their heritage.

GeoActivity Compare London's Immigrant Populations

Students' charts should resemble the following:

IMMIGRANTS	PUSH FACTORS	PULL FACTORS
Indian	Displaced by division of British India	Stability Jobs in textiles, health care
Bangladeshi	Upheavals after division of British India	Jobs in textiles and restaurant industry
Irish	Poverty Labor shortages	Jobs in construction, media, finance, and health care
Jamaican	Hurricane devastated economy Poverty and violence	Labor shortage—available jobs Education opportunities

1. Each country had once been a British territory or a British colony. It was probably easier for people from these countries to get permission to immigrate to the United Kingdom. Many immigrants may have already spoken English.
2. The economy has been strong since World War II, or at least stronger than the economy in other countries.

SECTION 2 REVIEW AND ASSESSMENT

Vocabulary

1. C	**3.** G	**5.** D	**7.** F
2. H	**4.** B	**6.** A	**8.** E

Main Ideas

9. They could rebuild their economies faster when they worked together.
10. The EU's economy is the largest in the world.
11. executive, legislative, and judicial
12. The euro has led to increased tourism.
13. They agreed to raise taxes and reduce spending.
14. the Soviet Union
15. The population's average age is increasing.
16. Migration within Europe increased as a result.

Focus Skill: Make Inferences

17. Members of the Common Market wouldn't need to pay tariffs on goods that they traded with one another, and would trade more as a result.
18. With stable democracies, there is less likelihood of conflict, which

19. The euro has increased trade among nations using it; Eastern European countries likely want to enjoy similar economic benefits.

20. Poland was still supported by the Soviet Union, and the protesters likely knew that armed conflict against their government would have had tragic results.

21. Yugoslavia was made up of a number of different ethnic groups that did not share much in common, and may have been unable to reach agreements or compromises in a democracy.

22. These older people lived much of their lives under communism, and may have preferred the stability of the previous system. Younger people do not share these memories and may be more comfortable with the risks that are part of the transition to democracy.

23. As older workers retire, jobs become available, attracting people from around the world.

24. Possible response: Health care providers would require new workers, both because current workers are retiring and because older people require more medical care.

Synthesize: Answer the Essential Question
Europe has clearly benefited from unification, with increased trade and travel throughout the continent. Adoption of the Euro has helped many economies, but has also made these countries more interdependent, so that economically stronger countries become obligated to help others in crisis. The transition from communism has given people across Eastern Europe much greater political freedom, but has had uneven results on the region's economies.

SECTION 2 STANDARDIZED TEST PRACTICE

Multiple Choice

1. A	3. B	5. B	7. B	9. D
2. C	4. A	6. D	8. A	10. C

Constructed Response

11. No, immigration helped make up the difference, but did not fully close the gap, as these countries experienced negative population growth.

12. Immigrants tend to pay more in taxes than they require in government services, providing revenue to cover outgoing social security payments.

Extended Response

13. The global economic crisis may have increased tensions between immigrants and native Europeans, as more people would have been competing over fewer jobs.

14. There were about 8 million people originally from Turkey and the Arab World. There were still 2 million more people from Africa in 2006.

15. Africa and Turkey are much closer to Europe than South Asia or Latin America are. It is probably easier and more desirable for migrating people to move to countries closer to their homes, which explains why immigration from Latin America is much higher in North America than in Europe.

FORMAL ASSESSMENT

SECTION 1 QUIZ

1. C	3. B	5. C
2. A	4. D	6. A

7. Immigrants from places with mainly Muslim populations, such as Turkey, North Africa, and the Middle East, have brought their religion with them to Europe.

8. The two world wars made many people feel that life was uncertain and unpredictable. Literature written during and after the wars often contains characters who express this feeling.

SECTION 2 QUIZ

1. B	3. B	5. B
2. A	4. D	6. A

7. In some eastern European countries, new businesses have been slow to develop. The economies of certain eastern European countries have experienced unemployment and rising prices.

8. Immigrants and native citizens often compete for the same jobs. Immigrants sometimes enter Europe illegally. There is disagreement about whether immigrants should assimilate into Europe's cultures or continue to practice their own traditions.

EUROPE TODAY TEST A

Part 1: Multiple Choice

1. C	3. C	5. C	7. C	9. A
2. A	4. A	6. A	8. B	10. D

Part 2: Interpret Maps

11. C	12. A	13. D

14. Most of Turkey is in Asia, not Europe.

Part 3: Interpret Charts

15. A	16. A	17. D

18. The phone calls in 2006, no matter what the country, were fairly inexpensive, and all prices were around 2 euros. In 1997, phone calls were more expensive overall, and prices varied more from country to country.

Part 4: Document-Based Question
Constructed Response

19. They lack a voice to speak on their behalf and perhaps even lack certain rights. They seem to have fewer advantages than other minority groups.

20. The author is sympathetic to their situation.

21. The Roma have left Eastern Europe because they were the victims of prejudice and sought a better life.

22. The Roma are no better off living in more democratic countries

today than they were living in eastern European countries under communist rulers. In fact, they are worse off.

23. The Roma population in France is more than four times greater than it is in Germany.

24. Most of the Roma have migrated to Spain.

Extended Response Possible response:

25. European governments could create new laws to help protect the rights of the Roma. Job opportunities and better education would help them earn more money.

EUROPE TODAY TEST B

Part 1: Multiple Choice

1. C	**3.** B	**5.** A	**7.** C	**9.** D
2. D	**4.** C	**6.** A	**8.** A	**10.** A

Part 2: Interpret Maps

11. D	**12.** B	**13.** A	**14.** B

Part 3: Interpret Charts

15. A	**16.** A	**17.** B	**18.** D

Part 4: Document-Based Question
Constructed Response

19. for about 700 years

20. They belong to no particular nation and have no one to argue on their behalf. They have also lived in Europe longer than other minority groups.

21. No, migration has not improved life for the Roma.

22. very poor areas of Africa and India

23. Spain

24. The Roma population is quite low in each one.

Extended Response Possible response:

25. The Roma represent a large group of people living throughout many European countries, yet they have are the victims of poor treatment and tend to live in poverty.

⬡ ACKNOWLEDGMENTS

Text Acknowledgments

322: Excerpts from *The Illustrated Bhagavad Gita*, translated by Ranchor Prime. Copyright © 2003 by Godsfield Press, text © by Ranchor Prime. Reprinted by permission of Godsfield Press.

374: Excerpts from *The Analects of Confucius*, translated by Simon Leys. Copyright © 1997 by Pierre Ryckmans. Used by permission of W. W. Norton & Company, Inc.

468: Data from the International Union for Conservation of Nature (IUCN) Red List of Threatened Species by IUCN. Data copyright © 2008 by the IUCN Red List of Threatened Species. Reprinted by kind permission of IUCN.

⬡ National Geographic School Publishing

National Geographic School Publishing gratefully acknowledges the contributions of the following National Geographic Explorers to our program and to our planet:

Greg Anderson, National Geographic Fellow
Katey Walter Anthony, 2009 National Geographic Emerging Explorer
Ken Banks, 2010 National Geographic Emerging Explorer
Katy Croff Bell, 2006 National Geographic Emerging Explorer
Christina Conlee, National Geographic Grantee
Alexandra Cousteau, 2008 National Geographic Emerging Explorer
Thomas Taha Rassam (TH) Culhane, 2009 National Geographic Emerging Explorer
Jenny Daltry, 2005 National Geographic Emerging Explorer
Wade Davis, National Geographic Explorer-in-Residence
Sylvia Earle, National Geographic Explorer-in-Residence
Grace Gobbo, 2010 National Geographic Emerging Explorer
Beverly Goodman, 2009 National Geographic Emerging Explorer
David Harrison, National Geographic Fellow
Kristofer Helgen, 2009 National Geographic Emerging Explorer
Fredrik Hiebert, National Geographic Fellow
Zeb Hogan, National Geographic Fellow
Shafqat Hussain, 2009 National Geographic Emerging Explorer
Beverly and Dereck Joubert, National Geographic Explorers-in-Residence
Albert Lin, 2010 National Geographic Emerging Explorer
Elizabeth Kapu'uwailani Lindsey, National Geographic Fellow
Sam Meacham, National Geographic Grantee
Kakenya Ntaiya, 2010 National Geographic Emerging Explorer
Johan Reinhard, National Geographic Explorer-in-Residence
Enric Sala, National Geographic Explorer-in-Residence
Kira Salak, 2005 National Geographic Emerging Explorer
Katsufumi Sato, 2009 National Geographic Emerging Explorer
Cid Simoes and Paola Segura, 2008 National Geographic Emerging Explorers

Beth Shapiro, 2010 National Geographic Emerging Explorer
José Urteaga, 2010 National Geographic Emerging Explorer
Spencer Wells, National Geographic Explorer-in-Residence

Text Acknowledgments

RB107, RB112: Excerpt from *The Face of War* by Martha Gelhorn. Copyright © 1936, 1988 by Martha Gelhorn. Used by permission of Grove/Atlantic, Inc. and Aiken Alexander Associates Ltd., London.

RB119, RB124: Excerpt from "Europe's Roma: Hard Traveling" from The Economist, September 2, 2010. Copyright © 2010 by The Economist. Reprinted by permission of The Economist Newspaper Limited, London.

Photographic Credits

83 ©Nick Haslam / Alamy. 93 © Mary Evans Picture Library / Alamy. 95 ©Eye Ubiquitous / Alamy. 99 ©The Bridgeman Art Library. 103 ©The Bridgeman Art Library. 119 © Lebrecht Music and Arts Photo Library / Alamy. 129 ©mpworks/Alamy. RB23 (t) ©Ray Roberts/Alamy (b) ©Everett Collection Inc/Alamy. RB45 (tl) ©Bettmann/Corbis (cl) ©Bettmann/Corbis (cr) ©Corbis (tr) ©Bettmann/Corbis (bl) ©Bettmann/Corbis. RB48 (tl) ©SSPL/Getty Images (c) ©Bettmann/Corbis (b) ©Underwood & Underwood/Corbis (tr) ©Bettmann/Corbis. RB54 (l) ©Bettmann/Corbis (r) ©Bettmann/Corbis. RB71 (t) ©imagebroker/Alamy (c) ©Colin Palmer Photography/Alamy (b) © PjrTravel / Alamy. RB74 (t) © Bettmann/CORBIS (b) ©Hulton Archive/Stringer/Getty Images.

Map Credits

Mapping Specialists, LTD., Madison, WI.
National Geographic Maps, National Geographic Society

Illustrator Credit

Precision Graphics

Front Cover

©Christopher Chan/Getty Images